YEARS OF STRUGGLE
THE FARM DIARY OF ELMER G. POWERS, 1931-1936

THE FARM DIARY OF

COEDITED BY

The Iowa State University Press / Ames, Iowa / 1 9 7 6

YEARS OF STRUGGLE
ELMER G. POWERS, 1931-1936

H. ROGER GRANT AND L. EDWARD PURCELL

For **M.F.G., M.F.P., J.D.G.,** and **S.J.P.**

H. ROGER GRANT is Associate Professor of History at the University of Akron, Akron, Ohio. He holds the B.A. degree from Simpson College and the M.A. and Ph.D. degrees from the University of Missouri, Columbia. Grant specializes in American Western History and in American Political and Social History of the late nineteenth and early twentieth centuries. He has contributed journal articles to both state and national historical publications.

L. EDWARD PURCELL is Historical Editor for the State Historical Society of Iowa in Iowa City. He is editor of **The Palimpsest** and editor for book publications of that organization. He holds the B.A. degree from Simpson College and the M.A. degree from The University of Iowa. Purcell specializes in U.S. Recent Intellectual and Social History.

© 1976 The Iowa State University Press
Ames, Iowa 50010. All rights reserved

Composed and printed by
The Iowa State University Press

First edition, 1976
Second printing, 1977

Library of Congress Cataloging in Publication Data

Powers, Elmer G   1886-1942.
   Years of struggle.

   Includes index.
   1. Powers, Elmer G., 1886-1942. 2. Farmers—Correspondence, reminiscences, etc. 3. Farm life—Iowa. 4. Agriculture—Iowa. I. Title.
S417.P64A36   338.1'092'4 [B]   75-23308
ISBN 0-8138-0600-3

80-1284

CONTENTS

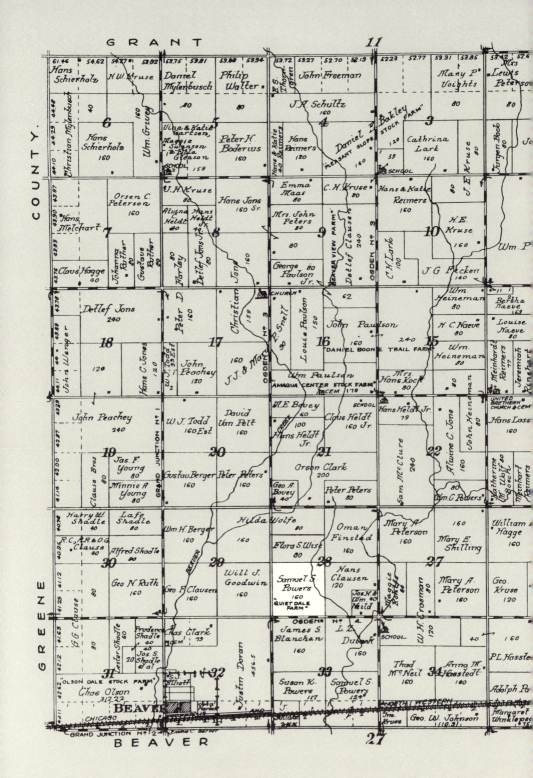

THE DEPRESSION DIARY of Elmer Gilbert Powers is a dramatic story of struggle and survival in the prosaic day-to-day record of farm life in Iowa during the 1930s. Living on rich land, perhaps some of the best in the world, Powers coped with meager prices for his products and fought the capricious whims of nature. His story as told in his diary is full, rich, and articulate. Elmer Powers's heritage—the land, simple education, religion—produced a conservative but concerned adult. He was almost a stereotype of the midwestern farmer.

Elmer's paternal grandfather, William Powers, settled in Amaqua Township, Boone County, Iowa, in 1875. A Pennsylvania native who previously had farmed in Carroll and Ogle counties in Illinois, he began an Iowa family tradition with his purchases of virgin prairie. Elmer's father, Samuel Powers, who in time inherited his father's property, bought two hundred and eighty acres of Amaqua Township land shortly after his marriage to a local girl, Susan Kate Gilbert. The couple specialized in raising purebred livestock; their long-range goal was to leave farms to their four surviving sons.[1] Elmer, the second oldest

AMAQUA TOWNSHIP, BOONE COUNTY, IOWA.
From **Atlas of Boone County, Iowa, 1918.** (Courtesy of The University of Iowa Libraries, Map Collection)

ELMER G. POWERS
AS A YOUNG MAN.

ELMER G. AND
MINNIE HARTLIEB POWERS,
WEDDING PICTURE.
APRIL 6, 1908.

surviving son, fell heir to the old family place. His brothers Sam, Walt, and Dan operated other farms in the area; like Elmer, they were tenants of their father until the patriarch's death in 1933.

Elmer was born on September 19, 1886. With other farm children he attended the neighborhood one-room country school. Since the nearby village of Beaver did not have its own high school until after World War I, Elmer and his brother Sam went to high school in Ogden, six miles to the east. Neither, however, graduated. Elmer continued his education through a variety of correspondence courses, notably from the Alexander Hamilton Institute of New York, and he spent a brief apprenticeship in Boone (the county seat) as a court reporter but soon returned to the farm with a healthy dislike of town life.

On April 6, 1908, Elmer married Minnie Wilma Hartlieb in the manse of the Presbyterian church in Boone. His bride, a native of Forest City, Iowa, had come to Beaver as a teenager, and there she met her future husband. Immediately after their wedding the couple moved into the old family homestead and set up housekeeping in the two-story frame house built by William Powers in the 1870s. Minnie gave birth to a son, Daniel Laverne, in 1912; a daughter, Lillian Lenore, was born in 1918.

Elmer had grown up in a strongly religious atmosphere. His parents, particularly his mother, were devout members of the pietistic Protestant sect, the Church of the Brethren, sometimes called the German Baptist Brethren or "Dunkards." Throughout his life Elmer attended church on a regular basis. In the 1920s and 1930s he supported the nondenominational Maple Grove country church near his farm home.

Politically, Elmer was faithful to the Republican party, a tradition of his family and community. A county history described his father as "an earnest republican,"[2] and local support for the Grand Old Party was strong. Amaqua Township voters, since the years of Republican ascendancy following the defeat of Bryanism in 1896, had cast the majority of their ballots for Republican candidates. Elmer remained in the fold of the party even through the depths of the Depression. Hard times, however, and the Hoover administration's inability to deal with the crisis caused Powers to tolerate the political heresy of the New Deal. Like other Iowa farmers, he found the

THE OLD HOMESTEAD OF WILLIAM POWERS,
BUILT IN 1877. E. G. AND MINNIE POWERS
APPEAR IN THIS PHOTO TAKEN ABOUT 1920.

cataclysm of collapsing commodity prices and drought to be politically searing experiences.

Although his religious and political sentiments give clues to Elmer Powers's personality, his writings and his children's reminiscences offer a more complete picture. Elmer was a meditative person, and the slow pace of horse-powered farming provided ample time for reflection. During the long, idle periods in the field waiting for the horses to rest, Elmer mused on day-to-day happenings. He also read a great deal and kept up on current affairs.

Elmer was not without prejudice—a strong antiurban bias permeated his diary. Farm people were the cream of American society in Elmer's eyes. He thought of "city folks" as less industrious, more careless, and more frivolous than "farm folks." Elmer himself fit his own agrarian stereotype by being hard-working, thrifty, kind, and gregarious. His idea of a spectator sport was a corn-husking contest, an activity he followed avidly. He enjoyed movies, but his principal form of recreation was attendance at community sales where he combined business and pleasure. He kept a close eye on the economic and social health of his community by gauging the mood of sales. Elmer always enjoyed trips to town where he could indulge his inclination to chat with friends and neighbors.

Throughout his life Elmer loved to tinker in his workshop. As his son recalls: "He went through a period, most of his life, as an inventor."[3] While Elmer never succeeded financially with this avocation, he developed several practical inventions designed to improve the overall quality of farm life. These included a milk strainer (to prevent spread of disease), a machine to candle eggs, and a mechanism to shock oats from a binder. Perhaps it was his creative streak that led him to dabble in cartooning throughout the 1920s. Several of his drawings appeared in local newspapers.

As a farmer Elmer Powers differed little from others in the community. According to census records the typical Amaqua Township farm in the early twentieth century contained a quarter section of land and was run by an Iowa-born owner-operator who grew corn and fed livestock. Only in the ownership category did Elmer differ from the norm. Yet he was not a typical tenant farmer since his father was the landlord, and Elmer apparently realized that eventually the farm would be his.

From the time he started to farm on his own in 1908 until the Great Depression, Elmer Powers's fortunes reflected state and local trends in agriculture. The "parity" years (1909-1914) and the World War I period saw not only high prices and a farm boom but also the development of farming as a highly complex business. Iowa farmers, continuing a movement begun in the nineteenth century, entered diversified farming on an increasingly

THE S. S. POWERS FAMILY, CIRCA 1915. At the back are Elmer's sister Elizabeth, brother Sam, Elmer and Minnie, and brother Walter ("Bill"). Seated in the middle are Samuel S. and Susan Kate Powers, Elmer's parents. In the front row are nephews Gerald, Paul, and Willard (sons of Sam); adopted sister Elnora; and Daniel L. as a baby.

mechanized and commercial basis. Livestock and cash grain crops became the staples of Iowa farm production.

These years were good to Elmer, and he responded by naming his operation "Quietdale Farm." He proudly proclaimed on his new letterhead stationery that he specialized in purebred Holstein-Friesian dairy cattle and Chester White hogs. However, the agricultural problems of the postwar era, which were destined to trouble all farmers, affected Powers. Although he successfully weathered the agricultural recession of 1921-1922, the continuing national problems of overproduction, lower prices, and increased foreign competition hurt him. At best, Elmer achieved only a modest prosperity during the 1920s, never getting more than a few hundred dollars ahead. Then came the drastic dislocations in prices and income following 1930. Resembling other midwestern farmers, Elmer Powers found himself in deepening financial trouble.

DURING one of his frequent outings to Des Moines in the twenties, Elmer struck up a friendship with the editorial staff of **Wallaces' Farmer and Iowa**

Homestead. Donald R. Murphy, one of the editors, urged Elmer to keep a "Day by Day on the Farm" account for the paper. "Some years ago," remembered Murphy in 1953, "I found a farmer in Boone county who liked to do extra things that weren't connected with farming and I started him on a farm diary."[4] The diary for Murphy and **Wallaces' Farmer** started in April 1931 and continued until shortly before Elmer's death on December 26, 1942. Only a small portion of the diary, however, appeared in the weekly publication.

The notion of a farm diary was not alien to Elmer Powers since he came from a family of diary keepers. His father and grandfather Powers both had noted the happenings of farm life for decades; in fact, "the family had the habit of just picking up any old book and starting a diary."[5] Elmer's first diary began before his marriage, and by 1920 he was typing a daily report, partly a chronicle of the day's events and partly a listing of income and expenses. By the late twenties his entries resembled those kept later for Donald Murphy, although they were less reflective and analytical.

Beginning in 1931 Elmer wrote several hundred words a day for **Wallaces' Farmer**, although he continued the practice of keeping a separate personal diary. For much of the 1930s he kept both the "public" and "private" accounts. The latter, however, generally contained brief notations. Moreover, these documents differed in two major respects; the private diary consisted of only a rough sketch of the day's events, devoid of interpretative remarks, and Elmer altered some of the personal names in the public version. For example, he called his brother Walt, "Bill," and referred to his children Dan and Lillian as "D.L." and "L.L." According to his son, he did this to insure the privacy of individuals who might object to personal references in a published account. Elmer consciously avoided the use of proper place names in the early years of the public diary; thus Ledges State Park became a "popular state park," and Boone, Ogden, and Beaver were called "county seat," "town," and "village" respectively.

Elmer Powers's ability to take shorthand and to type were invaluable skills for his massive diary-keeping projects. He frequently worked on his diary entries in the field as he allowed the horses to rest. Then in the evenings, based on notes taken earlier, he typed his entries on an old Oliver typewriter in a small office on the lower level of his farm home. In cold weather he wheeled the machine from room to room, seeking the warmest spot.

The results of Elmer's nightly labor were remarkable. Unlike most agricultural or "anonymous" diarists, his diary keeping was a highly literate enterprise. Elmer was an articulate writer, despite his lack of formal training, and he thought and felt deeply about issues affecting farmers and the country. Initially, there was a discernible strain of self-consciousness about the document. In the early months Elmer hoped that large portions of the work would be published in the farm paper, and he attempted to express the farmers' viewpoint to the general public. He was at times even coy or disingenuous. As the years wore on and little of the diary saw print, the self-consciousness melted away. Pseudonyms were discarded, and the writing style became more direct and less formal. Elmer occasionally expressed concern about his writing techniques: "I often wonder, when writing, if I should make an effort to write correctly. Or if I should write in the farm language and manner of speaking and expression of my own home community."[6]

"I should do more of my writing when I am not tired," Elmer admitted in November 1931. "Sometimes when I am tired, or when I have learned of a particularly unjust economic arrangement that is very unfair to the farm people, my pen just burns into the paper as I write. And the things I write are

A POWERS FAMILY PORTRAIT TAKEN IN 1918 WHEN BROTHER WALTER LEFT FOR SERVICE IN WORLD WAR I. (Identified left to right in relationship to Elmer.) Back row: Stella Kirkendall Powers (wife of Daniel) and brother Daniel, sister Elizabeth Powers Hutchins and husband Charles, Mamie Brown Powers (wife of Sam) holding baby Naomi, brother Sam, Minnie and Elmer. In front are: Sam's son Gerald, brother Walter, Sam's son Paul (in front of Walter), Father Samuel S. holding daughter Lillian ("L.L."), Daniel's son Oral, mother Susan Kate, Elnora (adopted sister), Sam's son Willard (in front of Elnora), and son Danny ("D.L.").

scorching." He acknowledged, however, "Invariably, later, I tear this up and write along an entirely different line."[7] Fortunately, he did not purge his entries of all critical thoughts. Indeed, the hundreds of entries Elmer Powers made in his public diary between 1931 and 1936 provide numerous historical insights. Most important are his accounts of the agricultural and national depression and his personal struggle for economic survival. Initially, Elmer was moderately optimistic about recovery. When others were feeling the lash of low prices and collapsing markets, he was isolated to a degree by his tenant status and the fact that he lived on rich land. As a farmer, Elmer was in perhaps the best position to weather hard times and he knew it. The earliest diary entries are notable for their frequent references to farming as a magnificent way of life. Elmer's paeans to agriculture began to wane only when the deepest crisis struck and in fact never completely disappeared from his writing.

Elmer Powers was hit hard during the early 1930s, and his problems are in his diary. The farmer's life—Elmer's idealizations to the contrary—was one of recurring crisis. Generally the crises were small; however, the crippling effect of rock-bottom prices left farmers no margin during the early thirties. Any crisis could be major under such circumstances. By late 1932 conditions worsened, and Elmer (and other Iowa farmers) found existence increasingly difficult. Ten cents a bushel corn, twelve-cent oats, and hogs at less than one-half cent a pound destroyed the farmers' last savings and credit. In January 1933 the price index of Iowa farm products stood at the lowest point in twenty-five years, only forty percent of the 1909-1914 average.

Elmer was shaken further in November 1933 when his father died, thrusting onto Elmer the financial management of his farm. When the father's estate was divided, Elmer received title to the farm on which he lived, but it was encumbered with a mortgage to an insurance company. Much of Elmer's time henceforth was spent in juggling crop sales and government checks to meet tax and mortgage payments.

As Elmer began to face management problems heretofore handled by his father, the importance of the New Deal agricultural relief programs became apparent. Elmer's diary entries chart his reaction to the changes in farm policy through the 1930s. Even though in normal times he was a political conservative, Elmer gladly accepted the massive movement of the federal government into agriculture. In fact, his criticisms of the New Deal were most frequently directed at local committees rather than the concept of federal intervention. Nonetheless, Elmer was disturbed in 1934 by the sight of idle corn acres that the New Deal removed from production. He realized that national surpluses had to be reduced and he needed the cash provided by the reduction program, but idle land went against a lifetime of experience. Elmer was perhaps at his best in discussing the corn-hog control section of the Agricultural Adjustment Act of May 1933. He explained clearly the complex and often bewildering features of this key New Deal agricultural program, from the hog contract sign-up to the final benefit payments.

The appointment of Democrat Henry A. Wallace as Secretary of Agriculture in 1933 pleased Elmer. He knew Wallace personally and greatly admired the former editor of "our favorite farm paper." Just after the 1932 election Elmer had written to Wallace: "Many of us farm folks have been wishing you could be our Secretary of Agriculture. This has been on our minds for a long time, regardless of politics and administrations."[8] During the Roosevelt administration, Powers continued to correspond with Secretary Wallace and generally defended his actions in the diary.

Despite his many problems, Elmer did not support the most militant Iowa farmers' movement, the Farmers' Holiday Association headed by the prairie radical Milo Reno. Elmer closely followed the activities of Reno, whom he had met in the 1920s, but found his brand of activism too extreme. Elmer even volunteered to act as a spy or agent provocateur for the Secretary of Agriculture by offering to join the Holiday Association. He wrote Wallace in 1933: ". . . if it will help me to get any information for you, or help you in any way, I will take a membership. . . . I will do anything you suggest or say to help run this thing down and to an end; we cannot let it go on or become worse."[9]

Even though Powers stopped short of Reno's extremism, he resented the blame placed on farmers for producing surpluses. In 1938, in a revealing comment, he wrote:

> . . . met a poultry dealer yesterday, a man trucking live poultry to Chicago. He says poor, hungry Chicago people come and take the sick and dead chickens to eat. Authorities try to prevent this. The man said half of the people in Chicago live from garbage cans. Many folks insist we have never, over a period of a year or two, had a surplus of our own production. Our economic and marketing system, with it[s] speculation, and imports of foreign goods, rather than our surpluses have been our trouble.[10]

If dislocation of the capitalist market system was not enough, weather in the Midwest dealt Elmer and his fellow agrarians what was for many the final blow. Elmer detailed the agony of the period from 1934 to 1936 when nature ran through one of her more brutal cycles. The major symptom was drought. Crops withered in the fields, livestock died from lack of water, and biblical-like

ELMER WITH TEAM IN FRONT OF THE BARN AT "QUIETDALE FARM," CIRCA 1920.

plagues of insects ravaged anything that survived the intense heat. Elmer was fortunate—his land was only on the edge of the worst drought area. The numbing, exhausting experience was vivid, however, in his day-to-day story. It is not difficult to imagine a community or a state on the edge of mass hysteria when Elmer speaks of seemingly green fields of corn in flames because of dryness or of the collapse of motorists who attempted to travel in the heat.

THE FARM DIARY of Elmer Powers came to the State Historical Society of Iowa in 1953 when Donald R. Murphy sent it to Dr. William J. Petersen, then superintendent of the Society. At the time, Murphy understood the historical value of the document. "It was always my intention," he wrote Petersen, "to send the material down to you, if you wanted it, on the theory that your successor some years hence might be interested in the diary of an Iowa farmer of the 1930s. . . ."[11]

The sheer bulk of the complete diary is overwhelming. The original contains nearly 2,500 typed pages, which cover the relatively short span of eleven years. Elmer seldom missed a day of typing, usually producing close to two hundred words for each entry. The document is complete except for some passages printed in **Wallaces' Farmer**. Editor Murphy ran extracts from Elmer's writings between 1931 and 1949; otherwise, the diary has never been published.[12]

An invaluable aid in preparing this edited version of Elmer's public diary was the parallel private one he kept from 1920 to 1942, which provides specific names and places. The original of the private diary (entries for 1934 to 1936 are apparently lost) is in the possession of Elmer's daughter, Mrs. Lillian Lenore Powers Gonder of Rippey, Iowa. A copy of the private diary has been donated by Mrs. Gonder to the State Historical Society and, along with the public diary and a transcript of a taped interview with Daniel L. Powers, is part of the Powers Papers collection.

When the State Historical Society compiled its **Guide to Manuscripts** in 1973, the Powers Diary came to the attention of the editors, and the entire document was read carefully. It was impossible not to become deeply involved in the life of Elmer and his family while reading this detailed chronicle of a difficult period in their lives. Despite the excitement of the diary, there were numerous repetitions and boring passages that accurately reflected the monotony of farm life. The recurring cycle of fieldwork and caring for livestock were unending themes in the story.

To reduce the diary to a publishable size, we chose to focus on the period of greatest drama and historical significance, 1931-1936, and prepared a brief epilogue for the years 1937-1942. Many of the descriptions of farm work, weather, and the routine of life have been removed from this edited version. Emphasis was placed on the most important events in Elmer's life. Hence the edited passages are somewhat more dramatic than the original. Even though Elmer tended to write entries of the same length from day to day, therefore producing almost an equal amount of material for each year, the less "interesting" years of 1931 to 1935 have been drastically reduced in bulk. Because the events of 1936 were especially dramatic—a severe winter, a burning drought, and pressing financial problems—we left most of that year intact.

In condensing the diary for publication, we refrained from shortening sentences and did not in general fragment paragraphs, although the length of some was reduced. Spelling and typing errors were corrected, a very few grammatical errors emended, and some punctuation altered to permit smoother reading; but we retained Elmer's use of simplified spelling of such words as "thru" and "bot." Notes have been provided to clarify various terms, persons, and events; and in some cases, such as with New Deal programs, rather extensive explanation proved necessary. Overall we strove to keep the flavor of Elmer's own writing.

H. ROGER GRANT
Akron, Ohio

L. EDWARD PURCELL
Iowa City, Iowa

ACKNOWLEDGMENTS

THE LIST of those to whom we owe gratitude is long and, as usual in such matters, we will be unable to acknowledge all our debts adequately. Most notable among those to whom we are obliged are Daniel L. Powers and Lillian Lenore Powers Gonder, the children of Elmer G. Powers. They graciously and patiently relived the days of their youth for us and allowed this publication of their father's diary. Dan's wife Dorothy and his son Dennis were likewise kind and helpful. Unless attributed otherwise, photographs have been provided by the Powers family. Colleagues in the History Department of the University of Akron and on the staff of the State Historical Society of Iowa were ever eager to help. Peter T. Harstad, Director of the State Historical Society, kindly allowed us to use the Powers Diary as we wished. Wallaces Farmer willingly permitted us to reproduce certain diary segments that had earlier appeared in their publication, and Gary Alban, editor of the Ogden Reporter, let us examine back files of his newspaper. The University of Akron provided a handsome research grant from the Faculty Research Committee, which was invaluable in our research. University secretaries Garnette Dorsey, Pat Godfrey, and Diane Markley typed drafts of the manuscript with skill and care. Finally, we must acknowledge a debt to our families who, in addition to the usual amount of neglect, put up with a summer "vacation" that consisted mainly of waiting for us to come home from the office.

H.R.G.
L.E.P.

YEARS OF STRUGGLE

THE FARM DIARY OF ELMER G. POWERS, 1931-1936

CHAPTER 1. **DAY BY DAY**

ON THE FARM: 1931-1932

Fri. Apr. 3. 31.

A good day to farm today. Did various things around the place, sawed some wood etc. The field work was plowing and disking with horses and tractor.

Tue. Apr. 7. 31.

We worked most of today in the back forty. I think it is one of the nicest fields I have ever been in; tho it does not now have any trees along the fences. The land lays higher than the rest of the farm and drains well. From this field, away in the distance, I can see the little cemetery where my Grand Parents and several of the others of the early pioneers are buried. The spot is marked on the landscape by a small clump of evergreen trees.

Wed. Apr. 22. 31.

This afternoon we drove into the village and on to the County Seat.[1] Called at the Farm Bureau Office. Also Highway Engineer and learned that our road is to be gravelled this season. Had some work done on the car. Most of the family went to a movie. I bot a rat trap and some things to experiment with in the line of bait. Do not know what results, if any, I will be able to report.

Noticed this evening that the barn swallows have returned. Another one of our corn planting signs. I think this same pair of swallows return to our barn and the same nest year after year.

Fri. Apr. 24. 31.

We attended the monthly meeting of our Township Farm Bureau this evening. No strangers or outside speakers were present. Just a meeting of the good home farm folks. One of the occasions that develops sympathy, creates understanding and binds folks together to work for the good of the community, State and Nation. The young folks were in town for their music lesson while wife and I were at the Meeting.

**THE FARMSTEAD AT "QUIETDALE FARM"
FROM THE NORTH SIDE OF THE ORCHARD.
DECEMBER 12, 1918.**

Sat. Apr. 25. 31.

With the morning work out of the way, D.L.[2] took the tractor and went to get the baler. I hitched to the grain drill and drilled millet in the lower pasture. This is just an experiment. Expect to drill other seed later.

I always carry a mechanical pencil in the apron of my overalls (yes I wear them, sometimes to town) and I find that I have broken this one some way. Have been looking over my collection of pencils this evening. Some of them do not show much wear, while others are quite badly worn. Each one has a history all its own. From checking auto races to writing up by-laws etc. in Farm Organizations and Co-operatives. And obituaries for funerals. And the dozens of stubs from reporting Hundreds of Speeches and Sermons.

Tue. Apr. 28. 31.

By the time we went to the field this afternoon, the wind was blowing hard and the sky was full of dust. We were having one of the worst dust storms I have ever been in. Much of the time one could not see more than several rods away. Occasionally it would slacken up, but usually the heavier particles rolled or swept fiercely along and the lighter particles were carried as high as one could see.

Sat. May. 2. 31.

I drove the harrow all forenoon. This is field work that I like especially well. The soil is in good condition and this harrow is very efficient. The cart is quite comfortable to ride on, and all around it is a pleasant pastime. I like to walk behind the harrow occasionally. I like the feel of the mellow soil under my feet. Then too, one can get a better idea of the seed bed they are preparing if they will walk over it some.

Thur. May. 7. 31.

L.L.[3] takes the last of her eighth grade exams today and I drove in to town with her this morning. Then I planted corn for the rest of the day and finished for the season. That is, unless it should be necessary to replant some of it. I think this is about the earliest date we have ever finished the corn planting.

Now, as fast as the planting is finished, the big worry will be "Will it come" "Is the seed good" etc. I have been looking at our first planting already. It is sprouting.[4]

The heavy part of our field work is practically all finished and our horses are in very good shape. They have not been sick or out of condition for a moment. Have not even had a collar or harness sore. Of course the tractor helps very much, but the horses always went to work first and quit last. I always get a new buggy whip when I start to plant corn. Sometimes depreciation on the whip is one hundred percent the first week. This season the whip came thru in good shape.

Thur. May. 14. 31.

Was harrowing over the highest place on the farm. And from the harrow seat could count one hundred and thirty five farmsteads.[5] From the largest to the smallest and from the thriftiest to the shabbiest, knowing them all as I do, I do not see the need of an apology for any of them. There is "good and sufficient" reason in every case.

Mon. May. 18. 31.

The first planted corn is up and one can "row" it both ways across the field. It does not have much color yet. Looks rather yellow. It is coming up quite uneven. Lack of moisture some places in the field causes this.

In watching the season's field crops as they advance in growth, it causes one to wonder how the work will all be cared for. It is staggering, the amount of work the farm families will do during the next sixty days. With the corn cultivation coming on, the alfalfa hay to get in. The harvesting to do, and it appears now that most of the threshing (here) will be done during July.

Wed. May. 20. 31.

Drove into an adjoining County this forenoon.[6] This county was early famous for its better roads, good horses, Angus cattle, etc., and Nationally and Internationally known because of its corn. These farm folks in this county are doing quite well in adjusting themselves to the existing times. In fact they apparently will lead in the business recovery, as they have usually been leading in the other things. Hope to drive over there from time to time to see if I am guessing right in this.

Driving along the highway, I picked up a young man who said that he was from the east. He described conditions there, as he sees them, and commented on the bitterness of many of those people. Replying to his question about people here, I told him that the farm folks here always have something to eat and, of equal importance, we always have our minds occupied and our hands busy. Two things that help much to keep people contented. He admitted that if the "Industrial East" were given many of the things they are now asking for, they would soon find that changing conditions would prove this sort of relief to be exceedingly burdensome.

Tue. June. 2. 31.

Brother Bill[7] said to use his horse while we had him here, so I hitched him in the team on the cultivator. Was working in corn stalk ground, spring plowed. Plenty of trouble with stalks and lots of hard work to do good cultivating.

But not a bad job at that. There wasn't any foreman who had to be pleased, or to find fault with my work. I know I can work here tomorrow. The place won't be shut down.

It is fine to be able to go to work, or to stop work when you please. Of course nature is exacting with the farmer. However, at the same time, nature gives one an independence not obtainable anywhere else.

Thur. June. 4. 31.

This afternoon D.L. drove the truck with hogs and poultry to town.[8] The hogs sold for four cents and the poultry for eighteen cents per pound. This is much the lowest price I have ever received for hogs. And I think the lowest we have received for this grade of poultry.

Fri. June. 5. 31.

We had an early lunch and started to drive to the Capitol City.[9] Stopped along the way and had the grease changed in the differential of the car. Spent a pleasant half hour with the Farm Paper folks.[10] Then drove on down town and attended to some business matters. The weather did not seem to suit me; the city seemed to bother me and I could not tell anything about the weather there. D.L. was attending a show and the ladies were shopping. As I came out of one of the office buildings, I understood why the weather bothered me. It began raining hard and I had not been able to see it approaching.

Every time I have been to a city and get back home again the farm seems very good. I like the cities for business places but, to live, for me I must always, "Have one foot in the furrow and one hand on the plow."

Tue. June. 16. 31.

I read in the daily today about several hundred farmers withdrawing their memberships in a farm organization. Poor policy, seems to me, at this time. Like trading horses in the middle of the stream. Without making any comment on the organization, I am wondering what kind of farmer members they were, who will leave or allow someone to talk them into leaving an organization in a time like this.

Sat. June. 27. 31.

This hot dry weather is very discouraging. Our garden is drying up badly and we water it frequently. The potato field is quite dry. Not a particle of moisture in the soil. The hay we took in today was cured as hay should be; it looked like it was flash dried and was nearly white in color.

The weather is very bad for all live stock. Many horses are dead from the heat. At the beginning of the heat wave the rendering plants paid the farmers for these dead horses, then they quit paying for them, and finally when they were piling up on them so fast, refused to take them at all. Many of these horses were badly needed on the farms right now.

Mon. June. 29. 31.

A very hot day today. Many farmers went to the field shortly after midnight and cultivated corn or mowed hay until this morning. We handled our work the same as we usually do. Except that the horses were taken to the watering tank quite frequently during the day. Unless we are very crowded with work we do not work our horses as hard as the average farmer does. The tractor, of course, could continue to work steadily all day.

Everything is hot tonight. The air, all out of doors. The house and the furniture. My writing table is uncomfortably warm to my hands. How long will this terrible heat continue? The loss of human and live stock life is recorded daily, but the crop loss cannot be ascertained until the end of the season.

Wed. July. 1. 31.

As soon as I was back from town I went to the field and cultivated. It was very hot for the horses, but we worked them carefully and finished the cultivating and put away some of the machinery. We had a late dinner tho. Another rain storm formed east of us just before noon.

After dinner I went to look at the road and hurried back because a heavy storm was forming in the west. We were able to get all of the small poultry under cover before the storm reached here. It broke with a wind storm, and rain followed immediately. The wind quieted without doing much damage and a steady rain continued for several hours. One of the most timely rains we have ever received. It came too late to be of much relief to some of the crops, but I believe the majority of the fields are saved. I attended our School Board Meeting this evening and was reelected Township Secretary.

Thur. July. 9. 31.

The greater part of the harvest is over now. It has been a nerve racking, body tiring job for most of the farm folks. Men, women and children have all felt it. They look forward to it with apprehension, are relieved when it is over with and not a one of them would want to farm without it.

Wed. July. 15. 31.

Brother Dan phoned that they were ready to thresh and would start this afternoon.[11] I drove the horses in from the pasture, greased the wheels of the rack wagon and went to thresh. Dan had a small field of wheat and they threshed it first. I hauled the last load of wheat bundles. Most of this load, I gathered up scattered bundles around the field and along the road. The grain was quite dry and threshed good and the threshed wheat was clean and bright. The field yielded thirteen bushels per acre. It was spring wheat.

Tue. Aug. 11. 31.

Returning from the village I drove north about school business. Many school boards, from the big city schools and the consolidated schools down to the small one room country schools, are having their troubles now. Several school organizations near here are planning to use local coal this year. Our board is one of them. I stopped past a very well equipped mine today. The quality of the coal from this mine is very good, but the mine has not operated very extensively recently. From the mine I drove to the County Seat and from there to the State Agricultural College.[12] It is a pleasure to visit both the College and the city near by. Every one at both places is pleasant, courteous and helpful.

Fri. Aug. 28. 31.

We were up and to the barns the very earliest this morning of any morning of the season. The cows seemed to resent their being brot from the pasture so long before daylight. We had a load of green corn for feed ready from last night and were able to get all of the morning work out of the way quickly. With several clocks around the place and a watch for each of us, it just happened that they were all on the blink and we did not know what time we started to the State Fair[13] or when we arrived there. Anyway we were on the grounds the earliest in the morning we have ever been.

We like to attend one of the first days of the Fair. Everything is much cleaner and more agreeable than nearer the close of the Fair. I think the attendance was as large as the same day last year. At least I seemed to be able to get in the way of as many people as I usually do at the Fair. Children were everywhere. It was their day. To me, the Fair seemed worth the effort just for the children alone. Personally, I thot the Fair just a little better than any I ever attended. I believe the most interesting thing our family saw at the Fair was not a part of the Fair at all. It was someone's car parked near ours. A car splashed and plastered with fresh mud. It surely looked good to us. I used to detest mud, in the roads especially, but now it seems the best thing that could happen would be to get stuck in the mud again.

Fri. Sept. 4. 31.

We finished the evening work early because I wanted to do more school work at the home of the Chairman of the Board. The ladies went along to visit. Driving out of our gate at dusk we saw a big fire and hurried to it. It had been burning for some time. Hundreds of cars were parked along the road and a serious crowd of farm folks were watching the fire. They are all confronted with the same possibility. This fire was not two miles from the fire last Saturday night. A large basement barn burned this time. Also a fine hog barn and the tool shed and work shop. With the exception of several machines almost the entire farm equipment was lost. The farm truck was burned, but the car was saved. Gas engines, grinders, cream separator, harness and all small hand tools etc. were destroyed. Most of the equipment was new. One horse and several cows perished in the flames. The only good well on the farm was several feet from the barn and could not be approached.

Wed. Sept. 23. 31.

Drove away this evening to attend to school matters. It is time to pay the teachers their salary for this month. I did not learn much today about the progress of the cattle testing war.[14] My sympathy is always with the farmer, but the Governor [Dan Turner] evidently feels that he must enforce the law so far as possible. Going thru the corn field I inspected the first fifty ears of corn I came to and found forty-seven of them showing signs of damage from insects and disease.

Sat. Sept. 26. 31.

We drove into the city this evening. Not nearly so many people on the streets this evening. The Cattle War was the principal subject of discussion. Many city people seemed to insist any diseased cattle must be destroyed by the owner, the same as they would be required to do if it was something they had in town that was a menace to the public health.

Tue. Oct. 6. 31.

I was awakened several times last night by the sound of the rain on the roof and in the eave trough. Everything was quite wet and muddy this morning. I walked to the pasture to get the stock. The hogs are rooting the pasture again and we penned them in the hog barn and ringed some of them. Ringing hogs is hard disagreeable work. With the many devices to aid the farmer in this work, main strength and awkwardness is still the principal requirement to accomplish it. I spent the latter part of the forenoon husking corn for feed.

Sat. Oct. 10. 31.

I worked in the "back forty." It is from this field that I can see the cemetery where my grand parents and other pioneers are buried. While I was putting on my husking hook I looked over the corn fields, hundreds of acres of them over the countryside, without a down row started in them yet. And I said aloud to myself, yes this corn husking is our portion for several weeks to come.[15] Then I thot of the ruinously low price of corn and as I looked again in the direction of the cemetery and saw the white gravestones glistening in the fall afternoon sunshine, I thot of the many, many farm men and women who had worked hard and toiled early and late and who were going into earlier graves because of these circumstances over which they had no kind of control. I do not like to keep thinking of these things and occasionally writing of them, but they are hard, cruel, bitter facts, very much with us; have been with us too long now and the scars of these days are going to show on the farms of this community for years to come. And worse than that, on the souls of the farm folks almost forever.

Sat. Oct. 31. 31.

Looking back over the past seven or eight months, it does not seem possible that the changes that have taken place in agriculture can be real. How the people of a Nation can show so little appreciation to their fellow citizens who produce the two very most essential things for them (food and clothing) is beyond my comprehension. There is but little consolation in the fact that while Justice is slow, it is certainly sure.

Today is the end of another month. We continue to hope for many improvements in many ways during the next month.

Mon. Nov. 2. 31.

I have been looking over our corn fields and thot it best to get someone to help us with the husking. The County Agent and the Welfare workers could not direct us to anyone. A prominent merchant remarked that seven men were waiting at his store door this morning when he came down, and they all asked for breakfast, which he gave them.

Fri. Nov. 13. 31.

A beautiful white frost covered everything this morning. It stayed on the corn most of the forenoon making husking a very disagreeable job. We have been most fortunate this season so far, as we have had only a few of these frosty mornings. Late in the forenoon and this afternoon, the weather was unseasonably warm. This warm weather was of much benefit in that it dried the fields and corn and made husking much better.

We tuned in and listened to the National Husking Contest. This evening we have not yet learned who the winners are. During the past week I have heard several farmers say they were not interested in husking contests, they saw enough of husking in their own fields. They are farmers who have never attended a contest. Many people are of the opinion that the husking contests will become the greatest sporting event of the year.

Sat. Nov. 21. 31.

Ten days ago we fully expected to quite easily finish the husking by today. There is at least eleven acres to husk yet and weather conditions look very unfavorable for future work.

We went to a movie tonight. Jim[16] expressed a desire to go. It was a Western picture which he enjoyed very much. I thot President Hoover and the Thanksgiving Proclamation was quite the best.

Mon. Nov. 30. 31.

This evening D.L. and I attended a Farm Union Meeting in the village. The attendance was large enough to indicate that farmers are becoming interested in caring for their own affairs themselves. Members of other farm organizations were in attendance as well as Union members. Also many farmers who do not belong to any organization. The speakers were quite logical in their opinions and very forceful in their statements of them and the audience applauded frequently and very generously.

Sun. Dec. 6. 31.

I did not do much writing or desk work today. I spent quite a little time looking over old farm records. In one filing cabinet I found some material I had put away during the epidemic of Hoof and Mouth disease some years ago. Time gives value to almost any sort of records.

Talking on the phone with a friend today I learned that I have been elected Farm Bureau President for our Township for the next year. If this is true, I think it is a big mistake on the part of the Township Organization.

Thur. Dec. 31. 31.

The rain that began some time during the night, night before last, still continued this morning. This is about the longest continuous rain that I can remember of. About the middle of the forenoon the weather became colder and the rain turned to sleet and everything was soon coated with ice. The trees and the telephone wires were soon in bad shape. This is one of the worst ice storms we have ever had. Soon after noon, the sleet turned to snow. As the ground is not frozen this may make the fields and roads very difficult to move about on. The radio kept us well supplied with advance weather information and we have everything in good shape for this particular kind of weather but the turkeys. They were the only living things on the farm that were outside. Before evening they were so wet and covered with ice that we easily drove them into their house. This is the first time many of them were in the house since they were poults.

I put in quite a bit of time today working on a summary of the farming business here last year. When I learned to be a Cartoonist, I learned to be a "futurist." All Political and Economical cartoonists must look well into the future. Now I have been trying to apply this to our farming operations and a year ago I attempted to anticipate the conditions as they might come up thru the year. Our final figures show that I have succeeded quite well and the farm earnings are almost as large, because of the many times I changed plans thru the year, as they were last year. But I am not going to say that I can continue to change plans fast enough during the next year to do as well as I have done this past year. I expect a year of terrific losses this coming year.

Fri. Jan. 1. 32.

Today is the beginning of a new year. I believe everyone is speculating more than usual as to what the New Year will bring to themselves and to all of us. Personally I still think the farm is by far the best place of all. The future may not look so good from a financial standpoint. However, for many folks the farm is more than a business and a place to try to accumulate wealth. It is life itself. First of all the soil, the feel of the earth. The respect they have for it. The fields. The weather and the changing seasons. All life itself comes from these several things. Then there is the plant life. The crops. The trees. The live stock and poultry and all of their young things to be cared for. The responsibility of growing the food and flesh for a distant and often unappreciative city. Just to be close to and work with nature is one of life's greatest opportunities.

Fri. Feb. 19. 32.

The weather was quite cold this morning but moderated during the day. We wished to move the baler to another farm several miles away today and we had some trouble to get the tractor started this morning. Then when we moved it away from the stack where we baled yesterday we found it would be necessary to take it across a plowed field which was not frozen solid. By hitching a team of horses on ahead of the tractor we moved the baler from the field to the main road. Neither the horses or tractor seem to work hard and it made a very good power combination. However, it was noon by the time we were out on the road.

This afternoon I took the car and left D.L. at the baler to move it on to the next job and I went to collect for other baling work we had done. Then I drove to overtake D.L. and the machinery. I found the baler tipped over on the side in the middle of the road and the tongue broken out and the tractor uncoupled. I soon located D.L. at a farm house where he was phoning for help. Returning home I loaded the wagon with timbers and blocks and phoned Al [17] to come with any lifting jacks he could find. Several farmers passing by stopped too and after several hours with lifting jacks we had the machine righted up and moving again. Nothing broken but the tongue and the tool box broken off. Considering that the machine weighs over three tons we were very fortunate. It was dark this evening when we were all finally back home.

Sat. Feb. 27. 32.

During the past several weeks a number of Meetings have been held in this vicinity. They were largely attended by farmers. The idea seems to be for the farmers to go on a selling strike. Whatever the results of these meetings might be, it will at least show how some of the farm folks feel and think about some of their many problems. I have not been able to attend any of these meetings.

Wed. Mar. 2. 32.

Today was moving day for many people and it was a damp cloudy day too.[18] Much better than yesterday tho. Twelve families in our neighborhood moved today. We have never moved and do not know much about it. The roads were in terrible shape. Four horses were required on every wagon and trucks were stuck in the mud frequently.

Thur. Mar. 31. 32.

I went to the barns very early this morning. The ground was frozen some and it looked like a good morning to break corn stalks. After an early break-fast we took doubletrees and teams and went to the field where we had left the stalk breaking pole and hitched a team of each end of it. By eight thirty o'clock we had twenty acres finished and that was all we wished to do in this field. As the morning was warming up we came back to the barns and finished a few neglected chores. Our stalk breaking pole covers twelve rows and breaking stalks is the fastest work we do on the farm fields.[19]

Fri. Apr. 22. 32.

D.L. and I both plowed all day today. He with the tractor and I with the horses. The tractor travels just a little faster than the horses and occasionally I must turn my outfit out of the way of it. The plowing speed of the tractor is about the same as that of the horses, but the horses must be stopped, some days frequently, to rest a moment or to hitch a trace that comes unhitched while turning at a corner of the field.

The soil in this field is loose and mellow and turns nicely. I like the soil. I like to have it in my hands. I like the feel of it and I like to walk on it. It is soft and yielding. Many people are supposed to like and covet gold but I prefer just good clean earth.

Sat. May. 7. 32.

The rains during the night last night left everything very wet this morning. We plowed timothy sod this forenoon. We had been leaving this work for a time when it would be too wet to do any other field work. D.L. used the tractor and I used the horse plow. This sod plowed very well. Occasionally the plow would not scour, but altogether it was a very satisfactory forenoon's work. I always like to plow sod. The furrows are so clean and neat and the furrow wall straight and smooth.

Wed. May. 18. 32.

We had some bad luck with the poultry last night. Rats carried one hundred and twenty-five of the day old chicks under the brooder house. And we had thot it was closed tight. We moved the house and killed a rat that was under it. This is the only loss we ever had among the poultry from rats.

Sat. May. 28. 32.

We cultivated corn all afternoon and tonight we drove to the County Seat intending to attend a circus there, but as we did not like the looks of the circus we attended a movie instead. It seems to me that the crowds of people that throng the sidewalks of many of the smaller towns and cities these Saturday nights is a quieter, more sober crowd than they were a few months ago. I have been looking for some improvement in the business and economic situation, but it seems now that everyone must go thru with it. No one can escape. This will be difficult for many of the younger folks to do. The older ones have become more schooled in some of these difficult experiences.

Mon. June. 6. 32.

I went to my desk the first thing this morning and made a list of the names of the people I wished to vote for today. Pat[20] came over this morning and I hired him to work for me some of the time during vacation. He and I cultivated corn this forenoon. D.L. came to the field and used my cultivator while the lady of the farm and I went to vote in the Primary Election. Folks at the schoolhouse where we voted said many more people were voting than usually did. Pat and I finished cultivating this field and started in the S.E. field just before noon.

Mon. June. 13. 32.

D.L. drove to town with the truck, marketing hogs. We were paid $2.25 per cwt. for these hogs. Much the lowest price we ever have received for hogs. Five years ago today they were $6.20.

Wed. June. 15. 32.

The Political Convention now on is the center of absorbing interest among farm people. They have a "wonder what they will give us" rather than a "we are going to have" attitude in the matter. Grain men are talking eleven cents per bushel for new oats today. I have been much impressed with the beauty of many of the mornings we have been enjoying lately and just now the nights are wonderful, tonight especially.

Sat. June. 18. 32.

The whole family of us went to town. I had to see the depository bank about our Township school funds. I also went to see the implement dealer about binder twine. We will have to pay seven and one half cents per pound for twine this harvest. We drove on to the County Seat. There I found a seed dealer who would trade me sudan grass seed for some timothy seed I have left from last season. Timothy seed will be figured at two and one half cents and sudan grass seed at three cents per pound in this trade. New oats were quoted at eleven cents and No. 3 shelled corn at twenty cents per bushel. All of the garden seeds we might want to plant for a late garden are almost as high as they ever were.

Fri. July. 1. 32.

With the beginning of this month we are starting on the last half of this year. It does not seem possible that harvest time is so close to us.

Pat cultivated sod corn and finished it before noon. We cultivated the crop only three times this year. Very few farmers are cultivating more than three times this year. Our corn prospects have never been better. By noontime a very heavy wind was blowing and we decided that we would not work at the hay this afternoon. I took Pat home after dinner. Stopping to talk with a party along the road, I learned that all of the banks in an adjoining county were closed this morning.

Farm friends from the other side of the county drove in this afternoon and we spent most of the afternoon visiting with them. The rural Telephone Company called the lines today, offering special inducements to farmers who would pay their phone rents at once. One farmer is milking eight cows and his cream checks just keep him in chewing tobacco.

This evening our Township School Board held their regular July 1st meeting. I was reelected Secretary for the coming year. There is a great deal of annoying work connected with it. The Board reduced the tax levy again this year.

Sat. July. 9. 32.

After the morning work was finished I drove to the field of late oats and decided that we should harvest them today. D.L. and I took the binder to that field and began work there. We knew that it would require careful work to finish that field today. At noon we were still wondering if we could finish it. Pat shocked busily all forenoon and was done with the early oats by dinner time.

This afternoon we went to the field early after dinner and were thru with the cutting and shocking by five o'clock. The weather was cooler today and the horses were more comfortable. The grain was much taller than in the other fields. A heavy wind blowing this afternoon caused us some trouble at times. The binder bothered too. It would make a number of large bundles and then a number of smaller ones.

Our grain shocks up well. The shocks are taller than last year. I think they are thicker on the ground. I think the yield will be satisfactory. But the price is all wrong. Nature has done her part well. Just men in their management are blundering. This harvest has been one of the most satisfactory we have ever had.

Sat. July. 16. 32.

While we were eating dinner the phone rang a line call and when I listened the operator said that all of the stores and the three banks in our town were having a Holiday for a week. And we always thot our banks were in such good shape. A mass meeting was to be held this evening. All of the banks in the

ELMER G. POWERS, NORTH OF HOUSE, CIRCA 1930.

County Seat are closed too. D.L. and I drove to town for a few minutes. The hog buyer was taking in hogs and men were planning to ship cattle, otherwise things were very still.

As we had the tractor ready we pulled the threshing machine near the barn and "set" it. It was too near evening to run it any and we wished to go in to town anyway. I stopped at our market town and the others of the family went on to the County Seat town. They were having a mass meeting there. In both towns the people thot it would be better to support the banks than to have them go into the hands of a receiver. This evening our entire community is confronted with a situation entirely new to them. It is the same thing many communities find themselves in. And as usual the farm people are carrying on in a splendid manner.

Thur. July. 21. 32.

At home this evening I hauled out feed, started the light plant and worked at my neglected writing work.[21] Our daily paper has stopped and we are not renewing it promptly. As a matter of economy I am resharpening old razor blades and when I shave I use any kind of soap instead of shaving cream. Many farmers tell me they are doing the same way. Several of the younger men have asked me about the old fashioned blade razors so they will not have the expense of the blades. I suppose some one will start the idea of growing beards. The oats market is a cent lower today. The weather looks very much like we will have some rain tonight. Meadows and pastures are burned up and the corn is getting in terrible shape. People do not seem to mention these things. I suppose the banking situation quite completely occupies their minds at this time.

Fri. July. 22. 32.

Many threshing rigs are in operation. They average about five of them threshing and one moving on the road. Not nearly so many trucks are in use as former years and not so much grain going to market at threshing time. The yield per acre is higher than last year and quality is most excellent.

Mon. July. 25. 32.

We continued threshing at Bill's place this afternoon. One of Bill's fields is near the railroad and little "Bobbie"[22] went to the field with me. He said he was going to drive my team from shock to shock for me. Our horses are not accustomed to seeing the train pass and one of them became frightened and began to rear and as I was standing near their heads I was struck by their hoofs. Bobbie was pulling on the reins just like a man. It wouldn't be a threshing job without a small boy or two around all of the time.

The banks in our town are open again today but they are continuing the Holiday in the County Seat. One of the men threshing today told of a Camp

Meeting held in a pasture field and twenty-four hundred people attended one service.

Fri. Aug. 5. 32.

One tenant farmer paying cash rent desires to turn the entire crop to the landlord in exchange for the rent notes, losing his season's labor. The landlord would not receive enough for the crop to pay the taxes on the land. And so it goes. The driver of the cream truck carries the cash with him to make payments for cream and he always travels well armed, expecting to meet hold-up men.

Sat. Aug. 6. 32.

While we were eating dinner this noon we heard a line ring on the phone; listening we learned the Farmers Holiday or "strike" was called for next Monday morning to begin at five o'clock.[23]

Mon. Aug. 8. 32.

Today is the first day of the Holiday. Many farmers are very serious about it. All agree that it cannot make things much worse, and that something must be done.

Tue. Aug. 9. 32.

Today is the second day of the Holiday. Many farmers were watching for the cream and eggs trucks. Some of them did not drive over their routes. Many of them did tho. Nearly all of them found business as usual. Perhaps five percent of the farmers refused to sell. Many cream checks amount to only from twenty to sixty cents. Nearly all drivers are traveling armed. The speculative stock market is higher today. Grain is up from one to two cents. Live stock slightly higher. Some factories are resuming work in a light way.

About our farming: we went to the baler this morning and worked until the middle of the forenoon when we had finished the stack. One hundred and seventy bales in all. We received a check to the amount of $13.50 for it.

Wed. Aug. 10. 32.

Later in the afternoon we repaired the truck and drove it to the village. The papers and radio bring us the news that there is a little better tone to all of the markets. The quantity of cream and eggs produced on the average farm is so ridiculously small that any advance in price is much appreciated. It is very hard to tell much about the progress of the Holiday.

Fri. Aug. 12. 32.

When I awakened this morning a steady rain was falling. The long looked for and much needed rain was here.

A rainy day on the farm, aside from its value to the growing things, furnishes a very pleasant diversion for the farm folks. Our cream ticket today showed a price of nineteen cents.

Sat. Aug. 13. 32.

This afternoon the whole family, visitors included, drove to the County Seat. It was a farmers' afternoon in town today. Numbers of them were everywhere. All of them were discussing the Holiday movement. Nearly all of them agreeing that it was worthy of and should be receiving more support than it was getting.

In our County Seat town, as a result of the Farmers Holiday, the price of milk has been raised two cents per quart to the producers. One producer tells me this advance of two cents per quart will amount to over five hundred dollars per month more for him.

D.L. drove the spreader this forenoon, working from the poultry houses. This afternoon, while in town, he roamed around watching folks and listening to them talk. He admires the way some of the farm folks stick up for what they think is the right thing and keenly enjoys some of the arguments they get into.

Tue. Aug. 16. 32.

About the "Holiday." A big Meeting was held in an adjoining County today. The driver of our cream truck weighed our cream today so if it should be dumped along the road he would have the weight. Driving along the road today we met loaded milk trucks carrying guards. Also we passed waiting farmer "pickets."

Thur. Aug. 18. 32.

More talk of the Farmers Holiday or strike movement reaches us. Something should be done to improve conditions. I do not know much about the plans of the movement. I do not believe it is properly organized.

Fri. Aug. 19. 32.

I drove to the village a few minutes this morning and listened to talk of the strike. I returned to the farm in time to talk with the driver who drives our cream route. He told me of some of the "bootlegging" some of the striking farmers are doing with their produce.

This afternoon we drove to the County Seat. The lady of the farm marketed eggs, successfully "running the blockade" the striking farmers had placed on the highway. We were able to do it while pickets were arguing with the driver of a truck load of ear corn.

Sat. Aug. 20. 32.

Tonight we went to the County Seat. I learned that yesterday and today pickets have been working on our highways, stopping cars and trucks in an effort to prevent farm products from reaching the markets. I have not learned just how this will cause an advance in the price per pound or bushel etc. The effort is commendable but the methods are questionable.

Sun. Aug. 21. 32.

In conversation with a picnicker from the northern part of the State I learned that Farm Holiday organizers there are paid one dollar for each farmer they can get signed up for the movment. Some farmers have been taken in on a membership fee of fifty cents. Others without any payment of any kind. Many Holiday officers and speakers are urging the farmers to dig deeper and deeper into their pockets if the movement is to succeed.

Mon. Aug. 22. 32.

All kinds of rumors are afloat about the Holiday Movement. It is pretty well established that the leaders can hardly be called farmers. Memberships are free some places, others on up, some donations etc. One of their speakers received fifty dollars for one talk. The Secretary receives eight dollars per day and his wife four.

Tue. Aug. 23. 32.

The local news of the Holiday is that two truck loads of cream were dumped yesterday. Approximately three hundred dollars worth. A station is opened in town where striking farmers may leave their milk and cream to be given away. One of my near neighbors left a thirty dozen case of eggs there yesterday. Clerks in the stores say much produce is brot in the back doors of the stores.

Wed. Aug. 24. 32.

The Farm Holiday is proving very unsatisfactory all around. Three cars carrying only women and children and no farm produce were showered with sticks and stones. One business man driving a new car was stopped by pickets and allowed to drive on, was struck by a stone and the rear of the car dented. Cream was dumped by boys under the directions of older men. A group of anti-strike farmers called on the sheriff, who seemed to be indifferent in the matter, and stated they would take the law in their own hands and keep the roads open themselves. There are possibilities of serious trouble soon. I drove to the village this evening. I should have attended a cattle feeders meeting in town but have had enough for one day.

Sat. Aug. 27. 32.

The Chamber of Commerce met with the Strikers today and as a result picketing is discontinued and the farmers who are working the movement are out seeking to get sixty percent of the farmers to sign up with them. Many say this cannot be done in this county.

Mon. Aug. 29. 32.

Our corn is drying and maturing very rapidly. Driving along the road today we commented on the changing fields. Also on the few tractors that were working as compared to the horse outfits.

Locally the strike seems to be a thing of the past. I heard today that the new corn crop price is to be eleven cents. One farmer sold his half interest in forty acres of corn for one hundred and forty dollars. L.L. started her second year in High School today.

Fri. Sept. 2. 32.

This afternoon was a very warm one and care was required in working the horses not to injure them. Many farmers are expressing regret, not so much that the Holiday Movement is not being handled properly, as that farmers do not seem to be able to affect an efficient organization that will be suitable to all or a majority of them. Anyway we are all of us learning something and in time we may get some place.

Mon. Sept. 5. 32.

The truck that takes our cream usually drives in here at the place around eight o'clock in the morning. Today when I came from the field at eleven o'clock, it had not been here yet. The ladies here at the house said many of the folks were uneasy about what had happened to it, suspecting pickets were at work. However, just at noon the truck came. Engine trouble had delayed it. Many farm folks had begun to worry about it tho. The driver said several farmers had offered to help him if anyone tried to stop him.

This driver, a farm owner, a former Farm Union man, said a number of farmers came to his produce house a week after the beginning of the strike to talk with him about closing during the strike, and he told them he had to keep open to buy their produce when they had their hired man or a relative bring in the cream or eggs. Every farmer present admitted having sent produce to him. The driver said he told them they were not getting on with the strike because of their own Association members and not because people did not believe in it. He said he also discussed with them the unaccounted thousands of dollars that had gone into the hands of men now prominent in the Holiday movement. These farmers finally retired from his place and no one had called on him since that time.

Wed. Sept. 14. 32.

Sometimes I think these Depression Days are growing darker and longer. Unquestionably there are signs of an early winter this season and this would be a very great hardship to many people, both in the city and out in the country.

Mon. Sept. 19. 32.

Today is my birthday. By way of celebrating it I did not work very hard.
I shocked some fodder this forenoon and this afternoon I did a few things around
the farm. Every day I am certain to do some farming. Even when not in the
best of health I disregard the lesser aches and pains and the minor ills and have
some close contact with the live stock and the fields. Broken bones are probably the only thing that would keep me away from the barns.

Everyone is trading now. I did a little of it today myself, trading sorghum
for grapes.

As a matter of economy I shaved today with a dime store blade. One of the
smoothest shaves I have had for a long time. But it is the farm women who can
think out and do the economical things.

While I am writing this evening I hesitate and my mind goes back to some
of the birthdays that have gone by. It takes them all to make a lifetime.
Nothing can take the place of experience and nothing else can give one the
necessary confidence in the future. We will come out of these dark, troublesome days again, still richer in experience but at a tremendous cost.

Wed. Oct. 5. 32.

Everyone that I talked to today was much pleased with the President and
his speech last night.[24] Even several Democrats said they were going to vote
for him and many Republicans said they had been rather on the fence, but
knew now how they were going to vote.

Mon. Oct. 10. 32.

This evening we drove to visit with a neighbor. They will shell corn tomorrow. The entire last years crop. They have been working on the highway
some of the time during the fall months and thereby financed the farm. A
visitor from O'Brien County says they will soon begin corn husking there. They
have many machines but do not plan to use them this season. Help is cheap
and the corn down.

Wed. Oct. 12. 32.

This afternoon I drove in to town to see the bank and found them quite
ready to extend any reasonable favor to me or to any other farmer. Driving
to town I passed the first field of stalks this season. A machine had worked over
it. Early reports on yields indicate around fifty bushels to the acre. Some will
go much higher and others will not reach this yield.

Fri. Oct. 14. 32.

I drove away about school business this forenoon and when that was
attended to I continued to drive and look around. Conditions in some communities are simply appalling. Some farmers do not seem to have any intentions of
husking their corn. Others are going at it in a half-hearted way and a few are

working the same as in other years. Many folks intend to use some corn for fuel, others would not think of burning it. Some school men are considering corn as fuel. One court house is now being heated with corn. Altogether this trip was quite unsatisfactory. I came back to the place wondering what the future can have in store for many honest hard working farm folks.

Wed. Oct. 26. 32.

In politics many are saying that they do not have anything against Roosevelt, in fact they have much respect for him, but they are going to vote for Hoover because they are afraid to try a change just at this time.

Wed. Nov. 2. 32.

We husked corn this afternoon, that is, D.L. did. I knew when I went to the field that I might have trouble with my team and I did have. The sorrel is known as a "runaway" and the bay has the same inclinations. When they finally started to run while I was husking, I caught the bay by the bridle and would have finally stopped them if we would not have struck a barbed wire fence at an angle and I had to let them go and roll under the fence. After a time I located them and walked up to them. Twenty rods of fence were broken down, the tongue broken from the wagon and the harness pretty well ruined, but the horses were not injured.

Tue. Nov. 8. 32.

Today is Presidential Election Day. Last night during the night I awakened and found I was sitting up in bed. Something seemed to be wrong and when I turned my head toward the window I could see the ground was covered with snow. I thot perhaps it would just be a light fall because it was raining last night and the weather may have turned colder. I went back to sleep, but this morning a blizzard was raging. We did the morning work and talked about taking L.L. to school and then going to the polls to vote. The bus came on time and we delayed the trip to vote thinking that the weather would get better. At noon we thot it would surely clear in a short time and when we began to inquire about the main roads later in the afternoon we found it would be useless to try to get to our voting place.

This is the first Presidential Election I have ever missed.

Wed. Nov. 9. 32.

I was up until quite late last night listening to the Election returns as they came in on the radio. Everything seemed to point to a Democratic landslide and this morning those reports were confirmed.[25] Either the Republicans have failed to care for the affairs of the people and the Government or else they have saved a country from disaster and the people did not know it.

Mon. Nov. 21. 32.

This week is Thanksgiving Week. Many people think that they do not have anything to be thankful for. And they are confronted with conditions that are new to them and certainly are not to their liking. Nevertheless there are still many things that farm folks can be thankful for. There are supposed to be so many thousands of unemployed, mostly in the cities, and many of them are the folks that a few years ago left the farms because they thot they were smart enough that they did not need to work the long hours that certain seasons of the year called for on the farm. And they didn't need to work for the small pay that the farmer got. They could have shorter hours and make bigger pay than the farm could give them and they went to the towns and the cities instead of acquiring a small farm or garden plot. They traded the security of the land for what has it turned out to be.

I dressed turkeys for the Thanksgiving trade. Bill husked corn all day.[26] The weather was quite warm and Bill says the fields are in fair shape again. Rumor says a prominent banker shot himself this morning. Supposedly too many complications to attempt to find a way out.

Wed. Nov. 23. 32.

Almost every day now some farmer finishes husking. And some of them have large acreages to husk yet. The banker, who attempted suicide Monday when the examiners came, is still living. And we thot our local banks were very sound. We are turning our cream separator by hand now since the weather is cold. It takes longer to get the engine started than it does to separate the milk by hand.

Wed. Dec. 14. 32.

After supper tonight I worked a long time at my desk. When I tired of checking over the farm's figures I began to clean up my desk (quite a large roll top affair) to have it ready for next year. And when I had tired of this I looked up the records on former depressions and tried to find some little consolation in going over them.

Thur. Dec. 16. 32.

Another of those very cold mornings this morning. I mention the weather very often because it is one of the factors that has so much to do with farm life.

I wrote out two hog pedigrees, then we started the truck and Jim[27] happened to come along by the place and he helped us to load two male hogs and we started to the community sale in an adjoining county.

Talking with the farmers at the sale (farmers from four counties) many of them seem to be changing from worrying about the depression and blaming things on Hoover to trying to find some way out of their own particular difficulty and the difficulties of farmers in general everywhere.

Sun. Dec. 18. 32.

About a year ago one of my near neighbors moved away into a tract of saw timber that he had bot. Today we decided to drive there to see them and to see Bill and his Father who had husked corn for us and who we knew were cutting wood near there. After what seemed to be a long drive over very questionable roads we found the location.

We found the neighbor at the very end of the road. He had built a new house and barn, right in [the] thick of excellent timber. Pehaps road improvements will make his saw mill readily accessible. Bill and his crowd were living in a very poor shack but were busy and getting out a good quantity of wood.

Mon. Dec. 19. 32.

While D.L. was away from the place I heated water and dressed a hog. As soon as it was cool enough to haul I loaded it in the car, caught a young gobbler and drove to the distant woods, as near as I could to Bill's cabin. From there I walked thru the woods until I located Bill. He and his small brother, pulling hand sleds, returned to the car with me and hauled their dressed pork in to their camp on the sleds. The gobbler they carried. This insures them a supply of meats for a time at least and a turkey for their Christmas dinner. Their old car is still setting in our barn yard. Very likely it will be all the pay we will ever get for these supplies. Anyway, what is one hog more or less these kind of times.

Today the oats market is reported higher than the ear corn market. Not many farm folks are talking about Christmas gifts.

Sun. Dec. 25. 32.

The young folks attended Christmas Services in town and reported that the attendance was very small. Our family enjoyed a Christmas Dinner with the old folks at their home in the village. The entire family was there with the exception of my oldest brother and his family. No gifts were exchanged by the grown folks. However, the smaller children were well remembered. Father's health is very poor and while we hope to have him with us for many more Christmases everyone is a little uneasy.

Late this afternoon before chore time I did a few odd jobs around the place. A habit I cannot seem to get over. I just must do a little fixing or changing every day. This Christmas is perhaps the most different one I ever experienced and can remember of.

Fri. Dec. 30. 32.

Most of my time in the evenings now is taken up with closing up the farm records. I very likely will not spend as much time looking over the records of the past year as I should, but will plunge into the New Year full of hope and expectation.

CHAPTER 2. **COMING**

ELMER, "D.L.," "L.L.," AND MINNIE.

Wed. Jan. 18. 33.

I still do not seem to be able to work out a plan of farming for the coming season that I am certain will produce results regardless of how market conditions will be at the end of the season. A new clover variety plant offers some possibilities. If it is adaptable to our conditions, it would take next year yet to get it into production enough to show much profit.

Thur. Jan. 19. 33.

Everything was very quiet around town. Some of the farm folks that I sometimes see in town were attending the State Farm Bureau Convention in Des Moines. Reviewing the present business situation it seems to me many people are just drifting, do not have any definite future plans or aims beyond present needs. Very probably when the average farm family have planned carefully and worked hard there are conditions away from the farm that they cannot hope to control and which offset their careful planning and hard work.

Fri. Jan. 20. 33.

Early this morning I phoned to the neighbor who has the straw to bale and he does not wish to work today, which leaves us to make any plans we wish. We did think about driving to the Capitol City; however, decided not to today. The Manager of the Shipping Association phoned and we can ship stock tomorrow if we wish.[1] Listening on the radio, learned that Governor Herring had issued a Proclamation tending to halt foreclosures, which should be gratifying news to many distressed people.[2]

The markets are lower again today. Sixteen cents for butterfat. Ten cents for eggs and two and one half cents for lard.

Tue. Jan. 24. 33.

Usually I am quite optimistic about everything and overlook many, many of the disagreeable things, especially the impositions that are practiced on the farm people. However, sometimes when I see them given an especially raw deal it makes my blood boil. Some of the very unfair things that are done to farm folks are because of neglect or ignorance, but some of them are deliberately and willfully vicious. Conceived, planned and carried out that way.

Sat. Feb. 25. 33.

In the mail today I received a bank check and because of the many bank closings I had D.L. go to town and cash this check at once. While D.L. was away I did more work with the seed corn and when he returned he said he had bot a truck load of cobs for a dollar and a half and we drove to get them. They will make fine kindlings for the kitchen stove this spring. At this price cobs are worth one fourth as much as ear corn.

Mon. Feb. 27. 33.

I went to the barns very early this morning and, when the regular things were done and many extra ones also to leave everything in good shape for us to be away all day, we drove to the Capitol City. It seemed a long time since I had been there and I very much appreciated meeting and talking with the Editors. Around the place here I had thot of all the things I wanted to say when I met and congratulated the New Secretary of Agriculture,[3] but when I saw him in his office today I could not think of a one of them. He has hard days ahead of him for he feels keenly the size of the job and the importance of the place at this critical time. But millions of farm folks have confidence in him and while they are toiling ceaselessly and enduring hardships they know he understands their plight and will help them if it is at all possible to do so.

Thur. Mar. 2. 33.

We finished boiling the first maple syrup of the season today. It is unusually good. We probably should have assembled sufficient equipment and made a large quantity of syrup. Thirty trees are producing at the rate of a gallon of syrup per day, and we have two hundred and fifty trees large enough for production.

Sat. Mar. 4. 33.

Coming in from the barns this morning I tuned in the radio and was surprised to learn about the bank situation. Spent some time contemplating the past events that had led up to this and then the probable outcome of it. Going back to the barns I worked a part of the forenoon with the live stock and went to the village. In the village I could not learn anything that would give me much satisfaction. I knew that all of our local banks were in good shape and the trouble must be farther on up the banking line some place. While I was in town I listened to the radio and learned all about the Inaugural Program.

Tonight we drove to the County Seat. I wanted to see how people were reacting to the financial situation and found them seemingly attempting to make the best of it, tho for some it will be a pretty hard job. Not nearly so

many people were on the streets and in the stores this evening. A very raw
chilly wind was blowing from the southeast this evening, making it very
disagreeable to be outside.

Mon. Mar. 6. 33.

People today were listening to every radio and reading every newspaper
in an effort to learn something more about the muddled conditions of our
banking and money affairs.[4] I have not found a farm family but what is
getting along in good shape yet, but many in town are not so fortunate.

Tue. Mar. 7. 33.

Near noon D.L. and I loaded a few articles in the truck and drove to the
Tuesday Market Day Sale. We found the usual crowd there, tho they were a
little late in arriving. The auctioneer announced that the days business must
all be handled on a cash basis. No horses were offered and the offering of
cattle and hogs was light. A fair line of machinery and miscellaneous articles
sold at usual prices. Several money scalpers and traders were in evidence
at settling up time but they were not patronized to any great extent. Several
men I talked to today were complaining about the new Administration but
the great majority of them were in favor of giving them every chance and
co-operating as far as possible.

Wed. Mar. 15. 33.

Today has been a very fine spring day. A few farmers have ventured
into the fields and are raking and burning corn stalks. I have not learned of
anyone working in the ground as yet. All eyes are still centered on Washington. Locally the banks are beginning to open again and in a few days we will
know which banks will be allowed to operate again.

Sun. Mar. 19. 33.

We did not have any of the stock outside today only for short intervals and
then they seemed to mind the cold so very much. Everything is covered up.
The turkeys were quite uncomfortable because of the ice that was clinging to
their feathers. This evening we drove them into one of the houses and closed
the door, the first time this season they have been penned. Banking, Farm
Relief etc. are all forgotten while we are wondering how severe the storm
will become and how long it will last and if the supply of feed will be sufficient
for everyone's needs.

Mon. Mar. 20. 33.

Today was a real winter day all day. A husky blizzard was raging and it
continued the day thru. The main roads were kept open. No live stock were
allowed in the fields. With one exception it is the heaviest snow fall of the
season. I drove our car to the mail box to mail a letter to the Secretary of

Agriculture and to take L.L. to the school bus and I got thru without any trouble.

Wed. Mar. 22. 33.

The two most interesting new events today, to me, were the change in the German Government[5] and the progress in our Congress on the farm relief bill. When a plan is finally worked out it may be a long discouraging task to get it functioning as it is intended to. There will very probably be some opposition and the longer time goes on the greater the opposition will become.

Sun. Mar. 26. 33.

Much interest is being manifested in the beer situation.[6] Many farm folks regard it as a "city folks" proposition. One farmer remarked that it reminded him of the Bible quotation "and Jesus Beheld a City and Wept over It."

Many farm folks are indignant about the so called "Road Houses" that are being built in many communities. They are never patronized by the local folks. And some people think everything about them is alright because a license fee or a tax has been paid by the proprietor.

Tue. Mar. 28. 33.

Finishing the morning work in good time we planned to drive to the Capitol City and arranged with Bill to come this noon and look after things.

We drove to the Capitol Building, or perhaps I should say the State House and spent the remainder of the forenoon there. We were in the House of Representatives and were rather put out about the way the tax revision and reduction programs, on which so many people had hoped for relief, were side-tracked while a commission or something from the east will tell us how much and how we shall pay our taxes. No one thinks they will tell us how to get the money to pay them tho. Our lunch in the basement of the building was much enjoyed by the whole family. We are very proud of our County Representative and the part he is taking in our State affairs. If all of the members were as fair, considerate, and sensible as he is, legislation would move swiftly and efficiently and much would be accomplished for all of us.[7]

Wed. Apr. 5. 33.

We were much longer doing the morning work this morning than we usually are. For some time we have been planning to drive to the college at Ames and see if we could get trial lots of seed wheat and flax. We made this drive this forenoon. Seeds we were most interested in were not obtainable. Meeting several men of the Farm Crops Department gave us much helpful information about new legume crops. We obtained some sorghum seed for our planting and will compare it with our regular varieties.

Around the college at Ames I thot I could notice that the student body was much more serious and quieter than any other time that I have been there.

Thur. Apr. 6. 33.

Today is our wedding anniversary and tonight, by way of a celebration, we drove to the city and attended a movie.

Many farm folks are disgusted because Congress has passed the beer bill before farm relief. Just at planting time the farm bill should have had attention and beer should have waited for it. At any other season of the year it would not have mattered so much which was first.

Fri. Apr. 28. 33.

Radio news this morning gave us information about the farmers' actions in the northwest part of the state. This is a very serious mistake. Evidently the leaders of these particular farmers have led them too far or lost control of them.[8]

We plowed again today, stopping long enough at noontime to grind more feed for the live stock and poultry. Also we marketed the last litter of pure bred pigs. If the weather is favorable tomorrow we should finish our plowing.

Late this afternoon a fleet of strange looking trucks passed our place. While they were some distance away and I could not see them very plainly. Yet I am certain they were loaded with troops and supplies and what has been a lark for some farmers is very likely to turn into serious trouble. Having contact with nature as the farmers do they should know that there are times when one must "bow to the powers that be" whatever they are.

Sat. Apr. 29. 33.

Radio news tells of more trouble among the farmers and law enforcement officials and soldiers taking charge of the affairs of several counties. We did some plowing today, but because of showers we were not able to finish the field as we had hoped to do. Between showers we did odd jobs around the place and in the farm shop and I caught up some neglected writing and bookkeeping.

Tonight we drove to town and I made numerous apologies to business friends that not all farmers were alike in their methods of expression and that while the farmers in our county are protesting they are business men and gentlemen.

Tue. May. 16. 33.

I mailed a letter to the Secretary today. Principally commenting on the hog situation in our community. Writing letters, or any kind of writing for that matter, is quite a chore for a farmer to attempt very much of.

Sat. May. 20. 33.

Writing tonight is rather a difficult matter for me. Today has been a windy day and the air was so full of dust that many times we could not see the sun. The earlier part of the forenoon dust began to blow and by noon it was terrible. This afternoon we had the worst dust storm I have ever seen. Our fields did not blow to speak of but many of the adjoining farms did. We did not disk up our fall plowing so fine this year and now it holds better.

Tue. May. 23. 33.

This afternoon we went to the field and I finished the planting and D.L. harrowed one planted field and part of another.

It is always a relief to me, and I suppose it is to all farmers, when the corn is planted. And I am always a little anxious, regardless of how good the seed may be, until it is all up and showing well. The small grain crops and the pasture and meadow grass is growing very fast these days. Live stock are doing well under these ideal pasture conditions. Not many farmers were doing any field work today. Some of them do not have anything pressing just now and others cannot get into their fields until drying weather.

Tue. June. 6. 33.

It was very dry and very hot in the fields today. I think the heat on the back fields of young corn plants was the most severe heat I ever experienced. The soil is very dry and the heat is extreme. A fair wind was blowing and the horses appreciated this very much as well as our taking them from the field to the water tank very frequently. Every day we think surely we will have a rain soon. It will be three weeks next Friday night since we had a rain of any benefit.

Tue. June. 13. 33.

Today has been another very dry day. Yesterday I mowed thru the growing oats and today when I tried to take up the oats they were too light and thin. The cows are holding up well in their milk production. The driver of the cream truck says that many herds are showing the effects of the drouth by a sharp drop in their output of milk. Among other things for tomorrow we must grind feed and take in hay.

Thur. June. 15. 33.

We are still hot, dry and dusty. We finished crossing the corn today. This is two cultivations without any rain on the fields. Some farmers continue to cultivate and others say they will wait for a rain before moving the soil again.

The dry weather is making it a difficult matter to care for the cows. We change them from pasture to pasture and try to keep them in the shade during the day. One neighbor has burned his roadside. The grass seemed to be dead and he burned it without mowing it first.

Thur. June. 22. 33.

Still hot and dry this morning. Not a cloud in sight in the sky. Last night was a hot night all night. Very hard to get any sleep. We made fly nets out of burlap sacks and put them on the horses this morning. The Red Ball salesman came.[9] We did not buy anything. Dan's family came to pick mulberries. The telephone line man came collecting. We are laying by corn in the field west of the grove. So far as I can see we are not accomplishing anything other than to say we are going over the corn. As I type this page this noon, perspiration is streaming from my face, work clothes are plastered with dust and sweat. One fine thing about it tho is that L.L. made a big freezer of ice cream and that cools one for the moment.

Fri. June. 23. 33.

Pat came this morning to help us with the cultivating but we did not get to work all forenoon because of a shower that came up about the middle of the forenoon and while it did not last all forenoon it wasn't worth while for us to go back to the field before noon. The growing things needed hours of a steady rain and needed it badly. However, this shower is of much benefit.

Sat. June. 24. 33.

Just as I drove in the lot at dinner time I saw the young horse "Prince," a three-year old, one of the sorrels that we are not working now, had become entangled in the wire fence. Before I could get to him he had broken the wire. All of our horses wear halters in the pasture and I soon had caught and let him in the barn. A glance showed that he was badly injured, perhaps ruined. A phone call brot the vet in a few minutes and we soon stopped the flow of blood. We have perhaps the finest team of young sorrels in the neighborhood and now one of them will be blemished if he is ever any good again. We have been careless or crowded for time and could have had the fences in better shape, but this particular place was in good order until the horse got into it.

Mon. June. 26. 33.

At noontime I led the injured horse from the barn and tied him to the hay-rack so that we would have plenty of room to work around him, but when D.L. went to spray him with fly chaser preparatory to removing the bandages, he kicked D.L. on the leg so that he can hardly get around. I removed the bandages from the horse and applied dressings as instructed by the vet. The wound looks bad. D.L. can't cultivate corn and L.L. said she would take a single row for the afternoon that the work might continue uninterrupted.

Tonight several locals of the Farmers Union and the Holiday Association held a big meeting in the community and we attended. I have never been actively associated with these farm organizations and went tonight because Rep. Gilchrest was to speak.[10] I found at a pleasant picnic place a large crowd of farm folks and many city labor, veterans, etc., people there. Gilchrest did not show up to talk, but a lengthy program was given without him. Many of the speakers said they were not farmers, or were not members of the Union or Holiday Association, or did not know much about farming. A number of farmers from Northwest Iowa were there and a few of them gave short talks about their jail experiences. Some members of the audience seemed to worship them and others remarked that they were jail birds.

Finally, after the crowd began to be uneasy for excitement, [Milo Reno] the state President of the Farmers Union began to speak. He found fault with and criticized every branch of the Government, Wall Street, the money grabbers and used a few names and expressions of his own origin. After some little time of this he began a personal attack on the Secretary of Agriculture becoming at times quite violent in his remarks urging all listeners to refuse to have any dealings with the Adjustment Act in any form.[11] The Union and Holiday folks hung on every word of this while the others of the audience appeared much amused. It was very late when the meeting broke up. I wouldn't have missed it for anything. It was a typical old fashioned Farm Union Meeting. The Program rambled along after a fashion and very likely these two organizations think they had a very good meeting and much was accomplished.

Mon. July. 10. 33.

Today was to be our last harvest day. And it was. We were in the field as soon this morning as the grain was dry enough to work in. There wasn't a very heavy dew but there was some. Bill came past the field and stopped for a few minutes. The weather was very warm. L.L. brot us ice water and fruit juices to drink. We are strictly temperate.

Wed. July. 19. 33.

For us another threshing time has come and gone. One of the poorest crops ever. We have followed good farming practices, better than the average neighbor has, and our yield has been discouraging. Two other machines in our neighborhood threshed all day today on oats that yielded fifteen bushels per acre. Market reports this evening carry the news that wheat prices are down twelve cents, corn six cents and oats fifteen cents today. Immense quantities of grain have been moving recently too.

Fri. Aug. 4. 33.

The horse with the injured leg continues to improve slowly and things are going on pretty much the same as usual. The past several nights have been very beautiful ones. The moon is full tonight. While city folks are wrestling with "codes" we country folks continue our regular and even way. [12]

Sat. Aug. 5. 33.

We drove to the County Seat this evening. Not so many farm folks were in town as usual. While the farms do not have a "code" as yet, many farm folks are discussing it with the city people. Some farm folks believe in it and others will have none of it. I suspect a few of the town folks would like to avoid any contact with it, if such a thing could be possible. A few farmers are beginning to complain that nothing has been done as yet for farmers.

Sat. Aug. 26. 33.

Tonight we drove to the county seat town of an adjoining county. Sometimes we hear someone mention the possibility of a peasant class in our country. Tonight I found plenty of evidence that we already have a peasant class. Ten years ago I knew this county quite well and the last year or two I have been suspicious about it. I have visited it on many occasions but I never saw the one class of people I was pretty sure existed until tonight. Their walk, facial expression, clothing, etc., told the story. They are becoming accustomed to their conditions and are going to get along. They are the class of people that if they were in a city would be entirely dependent on someone else for their living. Here they manage to exist by themselves and are going to make things better in the future. They certainly have been over some hard going.

Fri. Sept. 8. 33.

Just before noon I drove to the village and found Father had taken a sudden turn for the worst. Several times thru the day I was there and we are much worried about him. He still continues to ask about the crop and market conditions tho.

D.L. snapped a load of ear corn for feed today. In driving around the country I notice more corn wagons in the feed yards at this date than I have in former years. I would say that while we have a surplus of corn, it is mostly in the hands of the banks or landlords.

Sun. Sept. 10. 33.

Father's illness overshadows everything else and because of it we spend as much time as is possible with him. He has rallied a little since Friday and we are hopeful that we can keep him with us for a short time yet. This evening his condition had improved enough that very short services were held in his home.

Tonight tho, there is a feeling among the family that he has spent his last Sabbath with us. We are hopeful for a few more Sundays but the Doctors do not give us any encouragement from one day to another.

In the early days, as fast as the communities could get school houses built, Father assisted in starting Sunday Schools in them. Later then he worked for and contributed to the various rural churches of the several denominations in our community. Father has always been a deeply religious man but he has always been very quiet about it.

Tue. Sept. 12. 33.

This afternoon we were finally able to get the pigs delivered in town and and by this evening they are well on the way to Chicago. Four hundred head of pigs went out of the village today. Our pigs weighed sixty pounds each. Many of the lot weighed more than this. We marketed ten percent of our drove of pigs. I think this is about the same proportion from the various farms. Many farmers were around the truck while they were loading and they all regretted taking these fine pigs out of the lot at this time. Also many of these same farmers agreed that something must be done. In quite a few cases farmers were entirely out of money and will welcome the opportunity to cash in on this.[13]

Wed. Sept. 13. 33.

Driving to the village this evening we found Father's condition about the same, if anything he is some weaker. It seems that there isn't anything that can be done for him.

There is increasing activity about farm prices and how the N.R.A. will be a failure if the buying power of the farmer isn't stepped up. We knew that all along. If the buying power of the farmer would have been kept up several years ago it would have helped much for everyone. Farmers themselves are much to blame for present conditions. One cannot continue indefinitely increasing production with a diminishing market and expect prices to hold.

Sat. Sept. 16. 33.

D.L. went to the neighbor's and worked with the corn cutting this afternoon. I went to the village and received the check for pigs we had shipped last Tuesday. They averaged fifty pounds in weight and the market price was $8.25 which made it a very good sale. Of course there was the marketing expense to come out of that. But just now it looks like a good sale. How it will look after prices have advanced, if they do, is another matter. Many of the farmers were quite satisfied with their returns on these pig sales.

Thur. Sept. 21. 33.

Another consignment of Government pigs was to go from the village and we took five more from our herd. A number of farmers were in the village this morning. Some are in favor of the plan others much against it. These pigs go to Chicago by truck. This will make 650 altogether from our village. This would be the equivalent of ten carloads of mature hogs after a while. One farmer was sending pigs that he had bot from S. Dak. Some of the lot had died and altogether he will lose money on the deal.

Mon. Oct. 9. 33.
D.L. plowed in the back 40 nearly all day. I did various odd jobs around the farm this forenoon and this afternoon, the lady of the farm accompanying me, we took a party of High School students to a popular park and then drove on into the County Seat. Returning we brot the students home and then spent some time in the village.

This is the first time we have been away from the farm during the evening chore time for months. It seemed very odd and we felt entirely out of place. It isn't that we prefer the associations of the live stock to humans. But we really have something to do during the evening time and our town friends just stall around and try to act like they have something to do evenings. Farm life is very often hard and discouraging but it surely has its compensations.

Wed. Oct. 18. 33.
The papers yesterday and today contain much about the hog and corn plans.[14] I do not know just how our farmers will take to this idea. I would venture the guess that many of them do not fully realize just how much corn and pork we have on hand and almost every man of them would like to increase his production just as soon as the price advances a little.

Thur. Oct. 19. 33.
Papers today carry news of the farmer's strike called for Saturday. I think they will get a little more attention this time. If they will leave off picketing or interfering with any business and state their views in a serious way almost everyone will be for them.

Mon. Oct. 30. 33.
Father is in very poor health. He is very low. We were called to the village to his bedside very early this morning. Later in the morning he rallied somewhat and we returned to our farm duties. D.L. husked some corn today. I did various things, keeping near the phone.

Wed. Nov. 1. 33.
Excepting for several trips to the village to see Father, I husked corn all day today. Father is very low. However, he knows all of us. I am afraid tho we will never be able to talk with him again about our crop and live stock problems and it looks we will soon be forced to get along without the benefit

of his years of experience. Of course the much greater loss will be the loss of Father himself. He cannot last more than a day or two at most. We have temporarily lost all interest in the farm situation, politics, markets, etc.

Thur. Nov. 2. 33.

I spent the evening near Father. He is very weak and resting quietly. An expression of contentment and peace, in fact anticipation, was on his face. The long months of terrible suffering were over. All of the children and many of the grandchildren were with him. One family would gather near his bedside, then another. We finally came home for a little rest. The faithful nurse who had cared for him so patiently and carefully thru the many months remained with him, promising to call us as soon as there was any change. We are certain he will not be with us tomorrow.

Fri. Nov. 3. 33.

I was awakened soon after midnight by a phone call from the nurse and we hurried to the village and Father's bedside. We found him sinking gradually away. He had not taken any medicine for more than twelve hours and was quietly sinking away. The end came at one thirty this morning. Mother, all of the children and many of the grandchildren were with him. While we will all miss his presence, his helpful advice and the benefit of his years of experience and knowing how he suffered so long, we cannot selfishly wish him back.

Messages were sent to distant relatives in several states. The undertakers came to do their work and we came home for a little rest and breakfast. We are just beginning to realize the extent of our loss. I made a long drive this forenoon to take the nurse to her home. One of the hardest tasks I ever had was to try to thank the nurse for the careful care she had given Father, the care that helped to make his last months endurable.

This afternoon a casket was selected and many of the funeral arrangements were made. Several weeks ago Father instructed that when he was finally gone his body should be examined to note the extent of his diseases. The attending physician and an assistant did this work tonight. Three of we sons and a grandson (D.L.) observed the work. While it was a very disagreeable task, we were much satisfied after it was done. We know that all that could be done for him was done as the disease progressed. I will not make any comment other than stomach ulcers had turned to cancers.

Sun. Nov. 5. 33.

We buried Father today. The forenoon was a stormy, snowy one. I drove to an adjoining county to get the nurse who had cared for Father. She wished to attend the service.

There were so many of the relatives coming this morning and many of them so far that the ladies aid of the other village church arranged a dinner for the family in the basement of their church.

The services and burial were this afternoon. Father was a plain man and believed in a quiet plain form of worship and we tried to conduct his funeral with these thots in mind. The service, while sad, was also satisfying. A good man is gone. He had lived a good life and served his family and community well. What is there that anyone can complain about. I knew that Father had many friends but I do not know where all of them came from today.

The burial service was beautiful, impressive and comforting. We all returned to the village this evening. It is "Mothers" now and our thots will turn from caring for Father to extra care of Mother and respect for Father's memory.

Wed. Nov. 8. 33.

This evening we went to the village, to listen to the reading of Father's will. He leaves everything to Mother, this is as we expected and pleases us. Of course it will be some bother and worry to Mother, but it will also give her something to occupy her time.

Sat. Nov. 11. 33.

For some time we have been planning to drive to the Capitol City and today we got away. Nearing the city we passed the farm of the Secretary of Agriculture and talked about stopping to see him as we thot he might be there. But we did not wish to intrude.

I would have liked very much to have remained in the city to hear Secretary Wallace talk and possibly meet him this evening, but I just couldn't risk it. I did hear him on the radio tho and enjoyed his talk very much.

Driving along the roads today and observing farming conditions I could see great need for almost instant action in aiding agriculture. Most farm folks are hopeful and are trying very hard to carry on but it is quite a task.

Today is Armistice Day. It seemed to me it was not very generally observed. At least not as much as it should have been.

Mon. Nov. 20. 33.

I am very tired tonight. With enough help that we got 425 bushels of corn in the crib I was quite busy all day. Erecting temporary cribs and looking after the machinery and live stock and trucking an occasional load of corn to town so that I can pay the boys when we are thru made a rather hard day of it for me.

Then tonight I have been figuring up on the crop and it is a big money loser. Just now I am thinking about letting all of the help go tomorrow night and finishing the remaining acreage myself. It would stop some of the loss that way. If the wind a week ago would not have blown off some of the corn, I would not have hired this much help.

Weather indications tonight are for bad weather soon.

Thur. Nov. 23. 33.

Today is the day we have been working for so long. We finished the husking today. Almost every farm family is very much relieved when the crop is all in the crib. Our crop had been better than the average, but husking has been an expensive proposition. With the fields in bad shape we felt we should get it husked as soon as possible regardless of the expense. It is fine to have it done.

The fields and barnyard and horse stalls seem like a deserted place now. This evening I figured late on corn yields and yields by fields etc.

Sat. Nov. 25. 33.

D.L. had planned all summer that this fall he would send a turkey to the Secretary of Agriculture and today when we began to make arrangements our Post Master and Express Agent both advised against it.
So I wrote the Secretary trying to explain. I hope someone near Washington will see that he is supplied. He is certainly entitled to the best.

Wed. Nov. 29. 33.

A neighbor phoned and wished us to truck a load of hogs to a packing plant in an adjoining county. We did this for him and found a neat little plant crowded with business. Many farmers are so thoroly disgusted with the Chicago packers that they are avoiding that market as much as possible. While we were in this town we attended their regular Wednesday market day sale. This mid-week sale is made up of everything but live stock. We found a large crowd in attendance. A sort of holiday affair it seemed. Men, ladies, children and babies were there. The offering was everything from canary birds to all varieties of poultry. Fruit, vegetables, furniture, clothing and what have you.

Returning we stopped at a rural church for a few minutes to look in on a corn-hog meeting and found a very lively discussion going on. Tonight we attended one of these meetings in a nearby town and found a tired, worried audience who went away seemingly more confused than when they came but confident that their own township organization could see them thru. A very busy day today.

Wed. Dec. 13. 33.

Many farmers are much vexed with the public works arrangement.[15]
They see men getting forty and fifty cents per hour for six hours a day, five

days a week. Of course a farmer cannot pay those wages for those hours, so farm work is at a standstill excepting for what the farm folks can do themselves. Many had planned needed work and improvements that will not be done now. Also other farmers would jump at the chance of hours and time like these men are getting. Two farmers told me today that they had planned to pay their cash rent and other bills January first with hog money. One has eighty-five head the other a hundred. Last fall there was a chance they could make it. Not a chance now tho. I don't know just what they will do. Hogs can't advance soon enough to let them out.

Sat. Dec. 16. 33.

Today I saw numbers of C.W.A. men at work. Their working clothes are much better than the average farmer wears. I noticed the contrast very much. Sometimes I think there are many farm women who really do more work hour per hour than the men in the C.W.A. and the farm women put in about three times as many hours in a day. The paper today carries an item about 200 men beginning mosquito control soon.

Thur. Dec. 21. 33.

An incident occurred in our community today that worries me some. The Annual Meeting of the Greene County Farmers Union was held in Grand Junction today. Milo Reno was to be one of the speakers. I had planned to go and at the last minute decided to attend the sale instead. At the sale I heard a party say it is all fixed, he can't say anything. This evening I inquired about the meeting and learned that when Reno arrived in town the Mayor met him and told him just what he would be allowed to talk about. And what he could not talk about. The Mayor had thirty deputies there to see to things. The thirty deputies sat well in the front of the Legion Hall where the meeting was held. These thirty deputies were mostly C.W.A. men. The thing that worries some of us is that if a Union man cannot speak his mind perhaps the rest of us cannot either. It is generally conceded that the Mayor got his orders from someone higher up and of course using the C.W.A. men this way was not the Mayor's original idea either.

Fri. Dec. 22. 33.

Not so much happened here today. We dressed turkeys for the market and other buyers were here that we could not supply.

The township chairman of the corn-hog committee was here and wanted me to be one of the other two men to assist him. I want to do all I can to help but I will have to know more about what I am trying to talk my neighbors into than he could tell me.

This afternoon I went to the Farm Bureau office and did not learn any more. This evening I phoned the chairman and he has found two farmers to help him. They keep insisting that the farmers themselves will have all of the say on the corn-hog program but so far they haven't had anything to say about it.

I do not wish to always be "throwing cold water" but it is pretty certain that before things are over some branches of the New Deal will "stink to high heaven" and possibly a number of them will. I am not so old but I can remember of a number of times when things were done to help out the farmers and everybody got all excited and we had meetings and committees etc. and anybody can see where we are now. Permanent farm relief must come from the farms and on up and not from away someplace and finally down to the farms. This kind of organization gets top-heavy and fails.

Wed. Dec. 27. 33.

A Curtis man was here today.[16] Our Journal and Gentleman subs run for some time yet. D.L. traded him a turkey for the Post. This sub agent could tell of the many communities where he claimed he had sold almost every farm.

Tomorrow is live stock day at the community sale. Several neighbors are planning to take all of their hogs but a few brood sows there to sell. Practically going out of the hog business. Thoroly discouraged. One of the reasons we changed administrations was to get "Equality for Agriculture." Many of these things that I write are not my opinions but those that I hear.

Mon. Jan. 1. 34.

I wrote 34 without any trouble today. Usually I always forget and write the old year for several days before I finally remember. The New Year is with us, came in today. I thot one year ago today that with the passing of the old year things would be better for everyone. Perhaps it has. In a way the old year was about the toughest one I ever had. The serious illness and death of my Father during last year is one of the greatest losses I can ever experience. Of course one cannot keep his Father forever, but a few more years would certainly have meant much to me.

Regarding the possibilities of the New Year—I always think of things as affecting all of the farm folks rather than myself as an individual—I am certainly wishing things would change. Excepting for the loan value of corn, everything is much the lowest in price that it has ever been. And the things the farm folks buy are mainly higher. There is some concern among farm people as to who will pay the extra expense of all that is being done. There is still the possibility that other industries will make a go of it again and agriculture will be left to get along as best it can. This has been the way it has been working in former years.

Farm folks, as I talk to them, seem to be beginning the New Year with renewed hope but with much apprehension.

Tue. Jan. 2. 34.

We are beginning to hear much loose talk about the corn loans and perhaps some of it isn't so loose at that.[17] In many cases the loan money did not reach as far as it was hoped that it would. With the loan money gone there isn't much prospect until another crop is produced. One farmer took his loan money to the bank planning to pay the bank some and distribute the balance among creditors; however, the bank took all but twenty dollars and now six merchants are suing the farmer. They cannot see anything for them until another crop and they are not in shape to continue so indefinitely. Some farmers have about decided that their credit is gone since they took a loan. Advance payments on the corn-hog plan will hardly get to needy farmers by March first and some of them are finding themselves in a more difficult position than they were before.

Wed. Jan. 3. 34.

I spent a part of the morning listening on the radio. A party was talking from the office of the Revenue Department in Des Moines. His subject seemed to be the Processing Tax on pork.[18] He mentioned "PT" form so and so, and so and so, and so on. There seemed to be about a dozen forms and plans for figuring out the tax on the ham you gave the minister and the two pounds of sausage you traded to the store keeper for stockings for the baby or the lard you traded for toilet paper and so on. From what I know of some of the farmers they won't take kindly to all of this. A farmer reads in the papers and magazines and listens on the radio (he could put in all of his time doing this and wouldn't get it all) to all of the things that have been and are being done for him, and about how much better he is prospering now than he was a year ago. Then he goes out to work a little and reflect on how the whole thing just don't fit together at all. From the amount of radio advertising there is on the air now I am almost convinced that the other industries are off on a fair wave of business and prosperity and as usual the farmer is furnishing the food and clothing for the nation and doing it at the same old loss.

Wed. Jan. 10. 34.

The radio and newspapers give increasing information about the corn-hog plan and we are beginning to look for our blank contracts. If it were not for the seriousness of the situation there would be comedy in some of the things we hear on the radio about how much better off the farmer is than he was last year.

Thur. Jan. 11. 34.

Much of the time that I was around the sale pavilion this afternoon I was listening to the conversation carried on by the various farmers assembled there. One hears the State Legislature mentioned at only occasional intervals. The corn-hog program is the principal subject. Wherever you hear those words mentioned you will notice farmers edge closer in order to catch anything they might learn about the working of the plan. "No we don't get the paper anymore" is a statement I hear quite frequently. Earl May and his station seem to be the most dependable source of information.[19]

Mon. Jan. 15. 34.

I got up at three o'clock this morning and went to work making maps of each school district in our township, so that we may list each farm and farmers. When we had finished the barn work D. L. and I drove to the County Seat and I attended the corn-hog school held for the township committee men.[20] It was very interesting tho difficult work. A Professor Paddock and Jay Whitstone had charge of the school. After a short time out for lunch at noon I noticed a few did not return. By the close of day's work late this afternoon we were beginning to grasp the idea. For some of the farmers it was a slow job tho. Frequently you could hear some of them using a little profanity in a rather low voice. We are supposed to continue tomorrow. I remarked to D.L. when we were returning home this evening that I had paid well for training that was not as good as this and he remarked that before we were thru with it we might find that we had paid well for this. Perhaps he is not in sympathy with the movement to reduce corn and hogs this way.

Tue. Jan. 16. 34.

This morning I returned to the corn-hog school. This time I caught a ride with a neighbor. The attendance was slightly smaller again this morning and again at noon I thot I missed a few more. Today we worked with the question and answer book and the Administrative Rulings and planned for county and township meetings. Our township had seven men in this school, where from three to five was required.

Fri. Jan. 19. 34.

Four Twps. in our county were having a corn-hog school in town today and I attended. When I went in the hall above the bank I began to arrange the furniture for the meeting. Another farmer soon came in and we soon had things ready. When the county agent came I assisted him much of the time. Several new men were taking the work and this made it quite interesting for those of us who had already had some training. I left the meeting early to get home to grind feed. The weather was fine all day and we did not have any trouble with the grinding machinery.

Sat. Jan. 20. 34.

As it will be necessary for me to have Supporting Evidence on my hog sales this forenoon I made a long drive to see the Manager of the Sales firm and to secure the sales tickets on the hogs we have sold thru this sale the past two seasons.[21]

Mon. Jan. 22. 34.

After supper we attended a corn-hog meeting in the village. Almost every farmer from two townships attended. Several Farm Union and several Holiday men were the only farmers absent.

We men who had some training in the corn-hog work were much disappointed in the way this meeting was conducted. Attending farmers complained that they did not learn any of the things that were bothering them in making out their work sheets and arranging their papers. I was able to get two more S.E. on our past hog sales today.

Thur. Feb. 1. 34.

Today was the first sign-up day of the corn-hog campaign in our township. The meeting place was the village town hall. I went there early, was the first one of the committee there. Others soon came. I had taken things from our farm office thinking they would come in handy.

Our chairman, Art Muench is a very large man and always speaks quite loud. Others of our committee are Herb Clark, Hank Naeve, Bert Bakley, Fred Harten and Alex Doran. Our township was divided into three sections and we were to have three sign-up days.

Each member of the committee was very anxious to sign up anyone bringing in a contract. I did not attempt to do any signing until I could see how things were going. As soon as some contracts were completed I took charge and in looking them over found many errors. Some of them I mentioned to the committee members who had filled them, but I did not mention all of the errors at one time. I thot that would discourage them.

At noon I left the hall long enough for a quick lunch with my Mother and went back to a hard afternoon. Working out a system was quite a task and by this evening I had things going pretty well. I just had to develop a contract record system. Some farmers left their contracts with the committee, others took them out for various reasons and I listed them all. Fifteen contracts were completed today. Tomorrow will go better. By mid-afternoon I was a jack-of-all-trades. Outlining the maps and printing the names on the contracts gave the committee men the most trouble. The hall was a dusty place and the fresh outside air seemed very nice this evening when I rode home with Muench and Naeve.

Fri. Feb. 2. 34.

Today was ground hog day. And to our committee the second sign-up day. Things went better today. Twenty-five contracts were signed. I have made a list of all of the farmers names in our township and I placed a red letter "CL" by all of the names that have corn loans and a blue "CH" for the corn-hog contracts. We have around 130 farms in our township. 70 have loans.

The committee seemed to feel more free to call on me for assistance today. I was on my feet all day and on the jump most of the time. Some of the farmers had their work sheets completed, others did not. The supporting evidence on the hogs was the biggest problem for the greater number of farmers. I disregarded clerical errors, misspelled words etc. And I looked for dates, acres, hog counts and signatures. Sometimes returning contracts to committee men five times before they were complete. And I am not saying they are correct now. Occasionally I would hurry out and catch a farmer at his car as he was starting for home and bring him back to help the committee man. It is all very interesting, if hard, work. I was approached today by a farmer I do not know well, who wished to arrange with me to have his hog figures appear so and so. A large hog base is much desired sometimes. This farmer wanted to make me an offer and I was to come to his place and fix things up for him. I had been looking for something of this kind and immediately declined. Also I warned him about approaching any other member of the committee in like manner.

Sat. Feb. 3. 34.

Our third sign-up day and we certainly did a land office business and at that are running way behind the number we should have signed up. I phoned the county agent twice this morning for advice. A county committee man called this afternoon and brot us more supplies. He wasn't a bit of help to us and spent the time he was there visiting with his relatives from another township.

I am using a section from our filing cabinet and some of our loose leaf books etc. The committee are beginning to see the results of team work and system. Farmers from other townships tell us things seem to be going very smoothly with us. Hank is a very careful worker. Fred hurries too much. Herb seems to be the authority on contract provisions. Alex is always in hot water about something. The other members plod along. Sometimes they all get in a heated argument among themselves for a short time, then quiet down and go to work again. It is almost impossible to write [on] a sign-up day. Committee members who have been thru them know exactly how it has been.

We drove to town this evening and attended a movie. It was a very agreeable change.

Our committee is really a fine bunch of fellows and they have done well considering everything. Next week we must do some follow-up work.

Sun. Feb. 4. 34.

This evening I drove to Fred's and we fixed up my work sheet, contract etc. Our brood sow quota will be 12 head. Our corn-pig production 55 head. Our corn acreage 65 and our contract acres 16.

Mon. Feb. 5. 34.

The local committee was not working as a committee today and I did odd jobs around the farm this forenoon. After dinner D.L. trucked hogs to market and I drove to Herb's about corn-hog work, then to see the county agent.

The chairman phoned this evening that he wants us all to work tomorrow checking contracts. Thursday has been set for the last day. As I recall we have not had a rider attached to a contract and I do not look for anyone to ask for the early payment. Also so far only two have taken out more than the twenty percent reduction.

Tue. Feb. 6. 34.

We were back to the town hall on corn-hog work again today. Much of the time was spent going over the contracts getting them ready to send to the Farm Bureau Office.[22] Several farmers came in and completed new contracts. The list is steadily changing over to farms covered by contracts. There is beginning to be some talk of compelling all farmers to take the reduction.

Some members of the committee are beginning to complain about the amount of time it is taking. They feel they would like to be working on their farms more of the time. Four of the six days so far this month have been spent in this work.

Feb. 8-21. 34.

Many things have happened here on the place, and many things have happened in the corn-hog work during the past two weeks. I have spent practically all of the time at corn-hog work and have not been able to keep up any kind of writing other than the few notes that I have been able to make from time to time.

Our temporary committee composed of six to seven men has been fine to work with. Our township has around 130 farms and 20 landlords that we contact. Of the 130 farmers around 120 have been or will be signed up. Of the ten remaining farmers, only two have refused work sheets. Not any of them have been disagreeable or unpleasant about it. The clean-up campaign has required much work and much driving but it is pretty well done now.[23]

Tomorrow, Thursday the 22nd, I will take up things at home again and they will go on as before. The corn-hog program, as I have been connected with it, has taught me much. Volumes might be written about conditions right here in our township.

Fri. Feb. 23. 34.

This afternoon was our local corn-hog organization meeting.[24] It was held at our Center Schoolhouse. The meeting was called for one o'clock. It was one twenty when our temporary chairman called the meeting to order. I acted as Secretary. The chairman explained the purpose of the meeting. One of the committee read the Articles of the Association. Then the roll call of the 120 signers out of our 130 farmers in the township was called. I read the roll call and seventy-six responded. Several came in later.

The election of a permanent chairman, a vice-chairman and a third committee man were all conducted by ballot. Herbert Clark, one of the temporary committee members, was elected for chairman. He [is] supposed to be the best informed man in our township on the plan. Henry Naeve, also a temporary committee man, was the vice-chairman selected. His work on the temporary committee was just ordinary. Arthur Muench was elected for the third member. Muench is a very large man, farms quite extensively, is a good organizer and a man who goes right ahead and does the job whatever it happens to be. I could not understand why he was not elected chairman. All of these men live quite close together. No effort was made to elect men scattered over the township more. Completing the report of the meeting this evening is about the last of the work I will have to do. Of course I offered to continue any way I could help but as this committee is working for pay now I do not expect to be called. It has been a great experience. I would not have missed it for anything.

Sat. Mar. 24. 34.

This afternoon I drove to the village for quarter round to finish the kitchen floor. In the village I found the corn-hog committee were working and they insisted they should have supporting evidence on my packing sows that were produced in 1931.

This caused me to spend the remainder of the afternoon and much of the evening gathering this evidence.

I learned today that the committee had given me a corn yield of only 47 bu. We had taken out our best corn land, land that actually produces regularly from sixty-five to seventy bushels per acre. This would not look so bad but a neighbor has taken out his lowest producing land and they have allowed him exactly the same yield. Both fields are in corn stalks at the present time and anyone can see the condition of things. Personally I do not care about the inconsistency of the thing. But it will react against the program if the figures are published. The neighbor is supposed to have political influence.

Sun. Mar. 25. 34.

I have been away from the place much of the day. Mainly looking up corn-hog evidence. There is beginning to be much dissatisfaction about the program. However, this may quiet down somewhat if a little money gets out soon on the hog bonus and the corn benefits. But at the best, the contracted acres are going to become a very sore spot as the season advances.

Sat. Mar. 31. 34.

Tonight we drove to town. Many folks are making every sort of possible purchase these last few days of this month. This is to avoid the sales tax. At the start, this is going to be perhaps the most unpopular form of tax we have ever had. Opinions may change however. The recently adjourned, Special Session of the State Legislature is being complained about more than any other that I can remember of.[25] If you can get a farmer to talk frankly and freely to you, he will very likely tell you that he is annoyed and harassed by every device that scheming politicians and greedy other interests can devise. The great majority of farmers are very quiet. Some organizations say the farmers are satisfied. Those who say they are not are classed as radicals etc. Personally, I am just afraid we are nearer to peasantry, despite what has been done and what has not been done for the farmer, than we ever have been before.

Sun. Apr. 1. 34.

Today being Sunday, and also Easter, and a nice day our car was on the road much of the day. Our whole family were together for an Easter Sunday dinner. City company were with us too. These city folks come to visit with us occasionally from time to time, and we often compare conditions. The man of the family has been employed in a prominent factory, when it is in operation, for the past seven years. Recently this factory has been quite busy and all of the employees fairly well paid. Now, when these factory workers drive out in the country to buy eggs, meat, milk etc., farm folks refuse to sell to them; even with offers above the market. I mention this because I believe there is a widening gap between the relations of the city and country folks. And I take incidents of this kind as an indication of it.

Thur. Apr. 5. 34.

Our young pigs and our baby chicks have been showing by their condition that they were not being properly fed. This morning I drove the truck to town and talked the feed dealer into letting me have six bags of feed for the poultry and two for the hogs and he will wait until I get some corn-hog money and I can pay him for the feed then. Our farm feeds do not properly nourish our young stuff. I stopped at the produce house and listened to the produce people (former farm folks) tell just what they thot of the way the Government was mixing up things. I admitted that I had recently looked at the calendar to see just how long it would be until Dec. 1 and the corn-hog contracts would be expired. At the drug store where I bot a small quantity of needed medicine, I found the druggist in the same frame of mind as the produce people.

This afternoon I went to town to see a friendly and heretofore always accommodating banker about a twenty-five dollar loan until we would get corn-hog money. The banker said the Government did not want any money loaned on corn-hog collateral. Suggested that I see one of the Farm Credit Associations.[26] Admitted the young pigs and poultry would be dead or fully grown before a loan of this type would get around to us and wound up by saying that it seemed to be the Government's policy to kill some of the stock anyway. Might as well let it be now.

Sat. Apr. 14. 34.

Tonight we drove to the County Seat. A light rain fell occasionally. Something we are very much in need of. It is rather disturbing to think of how long it has been since we have had a real rain. Remembering other dry springs we can recall just how light the hay crop was those years.

Tonight I heard some talk that farmers were "sticking the hog-corn contracts in the stove" that 150 farmers in a meeting said if they didn't have the first payment by May 1st they would plant the contracted acres. We are sticking it out regardless. That is what we started out to do. Probably we will squirm sometimes, but we will stick.

Sun. Apr. 15. 34.

This afternoon we drove to a popular State Park.[27] There were many visitors and among them all I did not recognize any farm folks. Some liquor was in evidence. The first time I have ever noticed any there. As I see it, repeal made the drinking of many and various things more common and more in the open. Numerous ladies were smoking too. Farm folks are almost entirely free from these practices; at least they are in our community.

Sat. May. 5. 34.

No rain again today. One wonders how long everything can continue to get along without moisture. This drought here in the springtime is going to prove a very costly thing.

Wed. May. 9. 34.

Another hot, dry, windy day. Planted most of the day. Found the pliers that I lost yesterday. My desk and papers are covered with dust.

Sat. May. 12. 34.

Wonder of wonders, a fair shower of rain this morning, washed the dust from the grass. Moved poultry houses and planted corn. But could not finish. A nightmare of a week because of the dry weather.

DAN POWERS ("D.L.") CIRCA 1920.

Thur. May. 17. 34.

Odd jobs this A.M. To the shop in the village this afternoon. At home this evening. Our contracted acres do not add anything to the appearance of our landscape.[28] Especially since it is beginning to look as if we won't have much of a crop on the rest of the farm. We have never had a drought like this one during the growing season.

Wed. May. 23. 34.

There were a few clouds in the sky this morning and some indications of rain, but we did not receive any. I spent the forenoon rebuilding an old cultivator. Buying a new one is out of the question. With these crop prospects anyway.

The vet came at noon and said the cow we thot was a delayed reactor did not show any evidence of T.B. He perhaps has had much experience administering tuberculin and reading 72 hour results. But we have had more experience with follow up observations and decided we did not want the cow in our herd any longer. So we loaded her in the truck and drove to the packing plant and sold her for straight. Whatever slaughter tests indicate is not our affair. I did not write down any of the valuations we placed on the cows before we tested them and I do not remember what we put on this cow either. For some time we have not figured that we have a co-operating government, but a domineering one. We used to have a co-operating one. The young cow sold for

$1.50 and as she weighed 1000 pounds we received $15.00 for her. Of course we lose any State or Federal indemnities. But they would have been so low and so long coming that they would not have been of any value in helping us thru the dry difficult time just now and that will be with us for some time to come. Saw several farmers disking up their corn fields today. No feed of any kind anyplace.

Thur. May. 24. 34.

This morning I suggested to the farm wife that I drive over our school district and count the live stock and survey the pasture and feed supplies and report the result under oath to relief agencies. She, practical experienced farm woman that she is, insisted nothing would be gained by this. Any report would be lost in a maze of committees and final action would come from the estimates of city men riding over the community and looking at things from the highway. So I spent the forenoon cultivating corn. And very discouraging it was too. A dried out, burned up countryside greets the eye. Our own live stock is on half rations or less. No one wishes to borrow money anyplace and besides where could the feed be bot.

My own private opinion about the corn-hog program is that it is pretty well blown up. Many farmers say never again for me. With one or two exceptions every farmer in our community did his very best back last February to meet all requirements. And fully expected to have his first money long before this time. And now feels the delays have been caused by someone away from the farm. The bonus and benefit money is much needed. If the drought is not broken soon, sealed cribs will be freely opened and used. On the theory that the Government can't put us all in jail.[29]

Wed. May. 30. 34.

Today is Memorial Day. Today we had a new, sad and painful experience. For the first time we have a grave to decorate. Father's death last fall is the first one in our immediate family and this is our first Decoration Day in the real sense of the word. It was especially hard for Mother. For more than 50 years she and Father had lived near this cemetery, knowing all the time that it would be the last resting place for both of them and wondering who would be the first to go. Mother sometimes says that she is glad Father is gone and does not have to suffer thru these terrible times we are having and she often wishes she was with him now. We younger folks, while we are quite bewildered, would like to find a way out.

Thur. May. 31. 34.

Having heard many rumors about the release of the contracted acres for any use except corn planting we drove to the County Seat to learn from the county agent just what may be done. Our information was not quite satisfactory. Later reports may confirm desirable news.

Today finishes perhaps the most discouraging month of our long farming experience. And there have been times in years gone by that we thot things looked pretty bad. But they were not dry times during the growing season.

Sat. June. 2. 34.

I spent much of the forenoon driving to see if I could sell some of our livestock. We are bound to be short of feed for some long months and borrowing money to buy feed is quite a questionable thing to do.

Everywhere I went farmers were working to increase their diminishing water and feed supply. Much argument is indulged in as to the use of the contracted acres. Some claiming one thing some another. Others suggesting waiting a few days. The Secretary of Agriculture probably isn't very popular with farmers now and Mr. Tugwell even less so.[30]

Mon. June. 4. 34.

We were behind a month on our phone rent and the lineman disconnected it today. Perhaps many farm folks will do without many things soon. We will all be down to the barest necessities before the depression, Governmental experimental ideas and the drought have passed.

Fri. June. 22. 34.

Here on the place we cultivated corn and when we tried plowing with the gang plow,[31] on the contracted acres, we found the ground too dry and hard. Some folks refer to their land out of production as Government land. I do not like this expression. I always say contracted acres. We have too much "Government" in everything now. I am predicting the expression "Individualism" will be a popular one in the near future.

Sat. June. 30. 34.

This afternoon I attended a dispersal sale of a Holstein herd. A neighbor is working thru the system of "going bankrupt" and the sale is one of the results. There were some very good cows offered and several sold as high as $50.00. I heard several men remark that it was a very good sale considering the circumstances and especially considering conditions the past few days.

Sun. July. 8. 34.

I am afraid that I did not keep the Sabbath Day very properly today. Having been away so late last night I did not go to the barns until a late hour this morning. After finishing the morning work and a few neglected jobs I went with D.L. to the village.

There I met farmer friends and several who had been members of the temporary corn-hog committee. They are not very well pleased with some of the things that are being done by the County Association. Another neighbor is taking bankruptcy as the only way out. I do not know if I spelled the word correctly or not, but I do know that to many farm folks this is the meanest and ugliest word in the language. And it is a word that many of them yet may be compelled to use in their own business affairs.

Sun. Aug. 5. 34.

The afternoon had been very warm and while we were doing the evening chores we noticed small clouds gathering and a light rain fell. We did not give it much thot tho and while we were milking the cows a terrific wind and rain storm struck us. Rain fell in torrents. The barn timbers creaked in the heavy wind, we expected to see much damage occur. However, all that happened here at the place was the hayrack and wagon were blown over. As soon as we could get to the turkey colony houses in the fields we found many of the poults had sheltered in a nearby corn field and their heads were bloody from being pounded by the corn stalks. Sixteen of them were chilled, drowned or killed from some cause. The corn fields were laid low and whipped by the wind. The phone was out and driving around the neighborhood we saw many small buildings had been blown away. Also several barns were nearly wrecked.

Fri. Aug. 10. 34.

Corn prices have advanced to around seventy cents per bushel now. This results in many angry farmers. Those who have corn to sell are feeling pretty good. But those who have been feeding live stock and poultry feel that prices of these things are not in keeping with grain. If they had the feed they have used, to sell now, returns would be much higher. Also there is a feeling that speculators are taking from both the sellers and the buyers, I mean the farmers who have corn to sell and the farmers who must buy.

Fri. Aug. 17. 34.

The Chairman of the local corn-hog committee came to see me today and wanted me to bale a stack of straw. We are rather tired of baling and would prefer to be at home now. Also he wanted the job done different than any other way we have worked. Several farmers suggested that we would have to do his work or our corn-hog contracts would not be satisfactory to us. Others insisted that we refuse to bale for him, just to show him that he wasn't entirely in charge of the whole township on everything. I mention these things to show the frame of mind some of the farmers are in. We will very likely do his baling and everything will be alright all around.

Sun. Aug. 19. 34.

Our quiet community was much excited this morning. It seems that chicken thieves had been in the neighborhood last night and had stopped at several farms. They had been frightened away at each place. I am afraid thieving operations will increase in the country, and considering the

frame of mind of some of the farmers, something serious might happen to
the thieves.

Mon. Aug. 27. 34.

It is becoming quite common to call the mixture of weeds and grasses that
is being harvested and cut for hay on the stubble fields and on the contracted
acres "Roosevelt Hay." If there are many sandburs in it (and there usually is)
it is "Wallace Millet." Immense stacks of this stuff are springing up on almost
every farm. We have some of it mowed here and will care for it as soon as we
can. Perhaps it will do for bedding. Some of it is selling. I have not learned
any prices on it tho. We made 225 bales of it for neighbors today.

Sat. Sept. 15. 34.

Everything is quiet here and I think more farmers should be cutting corn.
The corn-hog checks have arrived in the county and that is getting all of the
attention at the moment.

We were in the village a short time this evening and as we were returning
to the place we remarked about how cold the air was and discussed the possibilities of an early frost. A heavy frost or a freeze now would be a calamity.

Sun. Sept. 16. 34.

I was much surprised and quite a bit worried this morning when I found the
ground was white with frost. Ice had formed on the poultry watering troughs. I
do not know how much of the state was hit by this frost, but I am afraid it did
much damage here for us.

Mon. Sept. 17. 34.

Our crops show the effects of yesterday morning's freeze very plainly. All
of our corn fields show it to some extent and our sorghum cane, our cane for
hay, our late oats hay and our late planted corn were more or less affected by it.

Thur. Sept. 20. 34.

The weather was very nice this morning and as we had planned to go to the
County Seat to get our corn-hog check, we drove away on an apparent pleasure
trip.

Our county committee has been paying two townships in the forenoon and
two in the afternoon. Many of our neighbors were in the office soon after we
were.

I was much disappointed in the amount of our corn-hog check. Several
days ago, with the assistance of some who should know, we had figured it out at
$170.40. The check I received this afternoon was for $121.20. I did not inquire as
to the difference. Several others mentioned theirs was a surprise too. I did not
enter into conversation with anyone, but heard several say this would be the

last time. Many mentioned the corn loan as a good thing but not the reduction plan.

Wed. Sept. 26. 34.

Last year we had a three year old horse badly injured in the wire fence. Today I sold him for five dollars. A horse that would have weighed a ton and been very useful here on the place or in the market. In spite of all the care we can take we must have some of these losses. Experience has shown that we can plan on losing one horse for each 160 acre farm each year.

Sometimes I think we are getting too many boards and commissions etc., piling up too much expense. All of it must be paid sometime. A pay as we go policy might be better in some ways.

Wed. Oct. 10. 34.

This afternoon we attended a community sale south of us. Much of the offering at this sale consisted of fruits and vegetables. All manner of root crops and similar garden stuff had been grown, cleaned, scrubbed and topped and much of it was of a very fine quality. Farm folks had worked long and hard to produce and prepare it and the money it would bring was desperately needed many places around the farm homes. Yet it sold for a mere pittance. City ladies in their fine gowns drew themselves away from the poorly dressed farm women but were glad to almost steal their farm produce.

The long drive home late this afternoon was not a very pleasant one for me. There are still many too many things to be righted and evened up before there is much fairness in our society.

The old farm that has sheltered myself and my family, as it has sheltered my Father and Grandfather before me, seemed almost a sacred place when I arrived at home tonight.

Wed. Oct. 24. 34.

This forenoon I attempted to select enough ears of corn that would be suitable for poultry feed. It was hard to do, and required much time. As I handled this corn I thot of the many splendid crops of this wonderful grain I had grown without giving it a thot. Now if I could just have some of the good corn again. Many of us will soon have a new respect for corn and before another crop is grown some folks will almost worship it. Only an unheard of mild winter can avert much hardship and suffering among the live stock in many sections. It seems that too many of the men who are working in various state and Governmental positions are not fully enough informed as to existing and future conditions and ways and methods of meeting the problems as they come up. I would say too much politics.

Tue. Oct. 30. 34.

Corn husking was our field work this forenoon. I might mention that corn husking is not discussed very much among farm folks. We meet, say how is the husking. The replies are usually, well we are husking. Any inquiries as to yield or quality brings the answer that "I don't know just how much it is making, I do not know how long the rows are" etc. As a matter of fact it is almost a flat failure as a crop, but it is never mentioned that way.

Wed. Oct. 31. 34.

This afternoon I went to the field for my first corn husking of the season. This is perhaps the poorest crop of corn I ever worked with. The yield is low and the quality poor.

Politicians are very numerous and as the campaign nears a close I am certain there are numbers of farm folks who are not fully informed as to the true issues and I am equally certain that if some urging isn't done, they won't all go and vote.

Tue. Nov. 6. 34.

Today is election day. We went early to the voting place and cast our ballots the way we feel that many of the rural folks should and will vote. That we expect the city and town votes to control the election is conceded by many farm folks.

Returning from the election we finished the husking in the back 40. This field of twenty acres yielded about eighteen bushels per acre. This afternoon we started to husk in a larger field and found the crop the poorest I ever worked in.

Wed. Nov. 7. 34.

A moment spent this morning checking up the election returns of last evening revealed results as we had planned to expect.[32] Business men everywhere will work just a little harder now to get and keep things going. The election has removed some of the uncertainty of things.

Corn husking occupied our time today. The field clears very rapidly and the crib fills almost not at all.

Mon. Nov. 12. 34.

Al[33] came to husk this morning and we finished the job at noon. Of all the crops I have gathered, this is the poorest one. Our cribs are almost empty and almost all of the feeding season is before us. I look at the contracted acres and wish I would have them in corn. They are the most productive acres on the farm. The committee cut them down on yield until this drought year they would have produced as much as the committee allowed. The thirty cents per bushel we receive on the yield they allowed won't possibly buy as much as this ground would have raised this year. Now I, like many others, must sell or almost give away hogs because I cannot feed them or buy feed for them.

Tue. Nov. 13. 34.

I spent some of the morning checking over our farming situation. The only conclusion that I could arrive at is the same one neighbors have arrived at. That some of the hogs must be sold. The corn-hog plan was intended to reduce the number of hogs and bushels of corn, so the price would be higher. And it did that. But the drought cut the corn yield much more. The price of market hogs is higher. But the lighter hogs are selling very low because so many farmers must sell them. A few farmers have substantial bank accounts or several thousand bushels of old corn and stand to make a lot of money. These few are the minority and not the majority.

I trucked four of our lightest pigs to the community sale. These pigs weighed eighty pounds each here at the place and they sold for $1.40 per head. If I would have hired them trucked I would not have had much left. All stock sold lower than usual. Veal calves sold for two for a quarter of a dollar. Not the best of course but good calves.

Trucks continue to haul corn south and southwest. Farmers say they are buying it with feeder loans money. Several admitted they bot anything with this money. Several others admitted that they did not expect to repay the Government. That they couldn't and didn't think the Government ever expected it.

I talked with and listened to several hundred farmers from eight or nine counties and came away from the sale wondering what it is going to be for a finish.

Fri. Nov. 16. 34.

This afternoon we drove to the County Seat. Many farmers were in and around town. But I did not talk with so many of them today. More and more the business men seem inclined to talk about conditions out in the country. One business man asked me if it were true that farmers were killing lighter hogs and smaller calves, rather than try to feed them or sell them at a loss. One story is that a farmer killed five sixty pound pigs yesterday because when he had paid the market expense he would not have anything left. Another farmer killed three calves for the same reason. I answered their question by asking one. "Who started the pig killing idea?"

Sat. Nov. 24. 34.

Of the very many Thanksgivings I can remember, this one is an outstanding one in the few things that we think we have to be thankful for.

Today, beginning the last month of this year we are hopeful that we can use it for a new start for next year and a new start at living and farm life.

Nov. 25-Dec. 1. 34.

I have been so busy with various things this past week that I just could not find any time for writing. The weather this week has changed from almost ideal fall weather to complete winter weather. And we like many others were not entirely prepared for winter. Much of the week was spent in trying to find some satisfactory solution for handling the farms that my Father left to us when he died a year ago. As is common there are mortgages on all of it. Some due, and others nearly so. I think we are working out a plan that will be completed in another several days or a week at most. It is quite a problem for all of us. Assuming the responsibility and the indebtedness of the farms is quite a load for we boys. And it will be a great relief for Mother if we are able to accomplish it.

CHAPTER 3.

MINNIE AND ELMER POWERS, CIRCA 1935.

Thur. Jan. 3. 35.

Last night we family folks gathered at Mother's home and planned about the settlement of Father's Estate. We were up until midnight. I am to have this farm where I have lived and worked so long. The details will be worked out later. I went to the county seat this afternoon. The weather changed and we hurried home to attend to the live stock.

Sun. Jan. 6. 35.

The weather being nice, I spent some time walking over the place. While I know every foot of it very well, prospective ownership of it is something new to me. There are a great many changes that I will want to make, but these will take time. It is a very old farm and has been in our family three generations. I am glad now that I have taken such good care of the land while it was the property of my Father. Now my first thot is about the level beautiful fields and how best to maintain and increase their fertility. Our live stock and the farm buildings of course come next for care and attention.

Tue. Jan. 8. 35.

We attended the community sale this afternoon. I did not notice anything out of the ordinary about this sale. But in conversation with other farmers I learned that at a recent farm sale eight head of ordinary farm cows sold for an average of $48.00 per head and at another farm sale Holstein cows brot a price of $55.00 per head. This would seem to indicate that live stock are looking up in price.[1]

Wed. Jan. 9. 35.

Today was another of those cloudy, dark days. Visibility was better than for several days tho. Several farmer friends have recently made unfavorable comments on the delay we are experiencing in receiving our second corn-hog checks. While I was in the county seat this forenoon I made an effort to learn something about the payment date but did not find out anything definite. Some folks out on the farms feel that these delays are unnecessary and that a better system might have been worked out. Just at the moment I would say that the program for this year of 1935 would not have a very heavy sign-up right now.

Thur. Jan. 17. 35.

I spent nearly the whole day at the county seat. We were working on the Estate matters. I fixed up the Federal Loan Application for this place today.[2] Dan is taking the home place and he made his application today too. I paid a fee of $16.00. I think he paid the same. We are a little uneasy about how we will come out in this venture. We both know the farms and how to farm them. There is only so much that we can do. The weather, the kind of crops and the markets we will happen to have will be the deciding factors in the matter.

Fri. Jan. 18. 35.

This afternoon we drove to the county seat again. I wished to have the loan payment dates placed at October and April. This way I plan to raise spring pigs to meet most of one payment and fall pigs to meet most of the other. The mail today brot information about the corn-hog plan for 1935. I do not see how I can take it. In order to meet the loan payments I must produce a given number of hogs to a given weight and this will require a given amount of bushels and acres of corn. My corn and hog benefits along with the crop this year would not have met any payments of any kind. Perhaps someone can explain just how the plan will take care of my case. Late tonight the weather is very disagreeable, snow, turning to rain.

Sun. Jan. 20. 35.

I am writing this, this evening. Today has been a very disagreeable one. The wind has been blowing hard all day. The cold is intense. This makes it very uncomfortable for the live stock. Our barn is quite warm, but there are some changes that I want to make as soon as I can and it will be some warmer for especially the calves. All of the feed in the field is covered with ice and when it finally melts off the feed will be much inferior to what it was before the storm.

Our water tanks are empty. The windmill could not turn yesterday because of the ice. And today I climbed to the top of the tall steel tower four times in an effort to get it started. By carrying bottles of hot water in my pockets and pouring them on the mechanism I got it thawed loose, but it was just too cumbersome and heavy to run, even in the high wind that was blowing. I think the top of a tall windmill is about the coldest place I have ever been. Especially when the thermometer is ten below zero. And our thermometer has been below zero all day today.

Weather conditions being as they were today did not leave much time for reading or rest. I hope the weather will improve tomorrow. This is very bad for the young stock and for the turkeys too.

Mon. Jan. 21. 35.

Clear and cold this morning. Thermometer below zero. Cold continued all day. Ice still over everything. I could not get the windmill to run and pumped by hand. I put the stock in the sun by the barn and they were warmed up some. The corn-hog meeting planned for tonight was postponed.

Thur. Jan. 24. 35.

The biggest incident of the day, today and for a long time, was our attendance at the regular Thursday live stock sale at a popular community sale. All stock sold well. Like old times. One cow selling for $74.50. Many cows selling around fifty to sixty dollars. It was interesting to watch the faces of many of the farmers in attendance in this sale. The facial lines brot on by the past few years of hard work, care, worry and in some cases by despair, relaxed a little today. Frequently I heard the expression "do you think it will last?" Many agreed that prices would still advance. Not many farmers have any cattle to sell, but they are feeling better anyway.

Jan. 25-26. 35.

We were at the county seat about land matters this afternoon. Corn-hog checks continue to dribble in. A flock of so-called Credit men and bill collectors are annoying many of the corn-hog check receivers. They are supposed to do this collecting for the local merchants. The receipt of the corn-hog checks by the farmers is much advertised by the Farm Bureau, mainly as one of their accomplishments, and word is broadcast everywhere that the farmers are getting this money. This one thing will keep many farmers from signing for the next year.

Mon. Feb. 4. 35.

D.L. and I were in the village a short time this forenoon. Today is sign-up day in the corn-hog program in our township and I wanted to hear more about the program for this year. I do not see just how I can make the loan payment on the limited corn acreage and the low hog base that I am allowed. Because conditions are changing with me I do not think that I should have to limp along on the basis of the smaller production that I had voluntarily gone into several years ago. Not many of our farmers were in to sign up this forenoon.

While I was in town I called at the office of the District Production Credit Corporation and was surprised at the number of farm folks who were there on the matter of loans to finance their coming season's operations. While our banks are supposed to have much money in them either it is not for farmers or they cannot offer satisfactory security.

Tue. Feb. 5. 35.

It is reported this evening that only one fourth of the farmers in our township have signed up for the corn-hog program. I do not know if this is true. Some of them will be a little late, but there will not be as many contracts as last year.

Sat. Feb. 16. 35.

This afternoon we ground a quantity of feed and trucked young cattle to the village to ship thru the Association to Chicago. This is the first stock we have shipped for some time and were glad of the opportunity to try the Chicago market again. Trucks have been doing all of the marketing recently.

Tonight we went back to our old custom of driving in to the county seat for the evening. I do not know if we will continue the practice or not. We had not been to town on Saturday night for some time. I was much surprised. The streets seemed to be deserted. Not nearly as many people were in town as I expected to find. I went to the implement store and found prices of many articles at figures that seemed to be too high. At least I cannot buy them and pay for them on the basis of past farming conditions. I cannot decide about signing up to corn-hog contract. With loan payments to make I would hate to depend on corn-hog money for even a part of them. The checks are too unreliable as to amounts and time of payments.[3]

Sun. Feb. 24. 35.

This afternoon the weather began to change, a light mist was falling and this later changed to a snow. The storm increased and turned into a regular blizzard. Our turkeys are always out in the open during all kinds of weather and when this storm became very severe we discussed the necessity of shelter for them. D.L. insisted that they had always taken care of themselves and that they would be in good shape and come thru it fine, but late in the evening we found that their feathers were becoming loaded with wet snow and ice and the heads covered with ice. It seemed that the best thing would be to get them in the big barn and the only way to do this was carry them in. We would make several trips thru the deep snow and the blizzard, then after coming to the house to melt the snow from our eyes and faces we would make several trips again. In this way we finally had all of the flock under cover. They are very hardy and very independent birds, but you could see that they appreciated the shelter of the big barn. Had we known the storm would be so severe, we could have driven them inside early in the afternoon.

Fri. Mar. 1. 35.

Today is the first of March. This is moving day for many farm folks. Some of them are moving to their own homes, newly purchased. Some are moving from what have been their own homes, that they have lost very often from no fault of their own. Others are and always have been tenant farmers. Our sympathy is much with those who have lost their homes. Our best wishes go to those who are moving on to their new homes. We are not in any of these classes.

We are staying where we have been for years, but we hope it will soon become our own home.

Apr. 1-7. 35.

This has been oats seeding week for us. All of our sowing, all of our disking them in and nearly all of our harrowing has been done this week. We have worked the fields more carefully than in former years. The weather has been windy and drying, tho there hasn't been much sunshine. Our acreage of oats is larger than we usually put oats to. This will be all of small grain we will have. We did not sow any wheat this year.

Thur. Apr. 18. 35.

I was plowing this forenoon and crossed the road to the mail box to get the mail. The daily paper told about the death of Editor John Thompson.[4] I have known Mr. Thompson for a long time. I have lost one of my best friends. And Agriculture has lost a good friend and counselor. While I did not see Mr. Thompson very often his page in the paper seemed such a part of him. Almost every time the paper came I would remark to my family that the page was just like him. Last State Fair time he rode up town with us and I remarked that he did not seem to be feeling so well as usual. The last time I was in his office he took up a small model of a clay silo from his desk and handed it to me and said "We have discussed silos very often, take this home with you and put it on your desk awhile," and I did so. And now I will leave it on my desk for some time at least. Since my Father's death John Thompson's page has been my closest contact for field, soil, crops and live stock help. I drove in to town tonight to try and call Editor Murphy on long distance to learn about the funeral but seemingly could not locate him.

Sun. Apr. 28. 35.

I guess there isn't so much to write about today. Not many if any, from our community attended the big meeting in Des Moines yesterday. And not many of our farm folks tuned their radios to hear the Holiday Meeting or Huey Long.[5] We farm folks are busy at this time of the year and some of us feel that there will always be someone to do the shouting and noisemaking for us. Perhaps it does help some.

Mon. Apr. 29. 35.

We stopped at the cold storage plant and learned about the new plan of storing our fresh killed meat in our own rented locker, where it is kept at zero or the proper temperature for storage and where we can get it anytime. We plan to use this service. It will cost a dollar a month or nine dollars per year. We must be careful to have the animal heat out of the carcass and to cut it in convenient pieces and wrap them separately, so we can get them out as needed.

Wed. May. 1. 35.

The Federal Land Bank Appraiser was here most of the afternoon. He is a pleasant man to meet and go over the farm with. Also he is very efficient in his work. He was particular to know about the soil and drainage and about if there were any weeds in the fields. And he was very certain about the size of the various farm buildings. I do not know just what appraisal he will make for us, but it has been a pleasure to meet and become acquainted with him.

May. 7-14. 35.

Of this week there have been five of the working days that I have been in the field planting. I do not recall that I have planted any whole days. I planted when the weather and other things would permit and when the horses were in shape for it. Two evenings I planted until dark. But for some cause I was not able to go to the field early in the morning. We will have close to 72 acres in corn this season and when I came in this evening there was only four or five acres left to plant tomorrow. I would [have] much preferred to have a larger acreage of corn, but the reduction program would not allow. I am not sure that we can meet our loan payments quite as easily this way as if we had more corn.

I have not been able to learn anything about the progress of the Federal Loan. I suppose and I am hopeful that it is coming along in good shape.

Fri. May. 17. 35.

When we were driving to town this morning we met two truck loads of relief men going to work. We had been working almost five hours before they started. Sometimes I think we will have to make our days still longer to meet all of the increased taxes and expenses we will have. Butter fat is down to 26 cents and oats to 36 cents. Only the most skillful farmer, with the best of luck, can meet production expense on these kind of prices. There is too much waste between the producer and the consumer.

Thur. May. 23. 35.

Our High School held their Commencement Exercises this evening. L.L. graduated tonight. A very pleasing program. We are very proud of her and of all of the class. The rural districts still produce about the finest young people we have.

Fri. May. 24. 35.

The fields are much too wet to work in today and we drove to the county seat this forenoon. D.L. is getting newly hatched turkey poults from the hatchery.

I wanted to see about our land loan and some of the things in connection with settling Father's Estate. The loan Secretary has finally received a letter asking for a little more information. It seems that there have been other rulings made since we started our loan applications.

Sat. May. 25. 35.

We kept the harrow going in the corn fields all day today. Not many farmers in our community have harrowed their corn fields this season. We are planning on harrowing all of ours. I guess I am still old fashioned in this respect. It does save time not to harrow, that is you have your harrowing time for something else. I am not so sure about the crop production tho. And I still think the practice leads to a weedy farm. However, tenant farmers do not find much inducement to keep the fields clean, and our community is beginning to show it too.

Tonight I went to a cheap little movie on a side street and there I found conditions that I did not think existed in our community. It seems that the Government had just as well provide entertainment along with everything else for even the very poor will find a way to have some kind of it.

Tue. June. 4. 35.

Cultivating corn most of the day. Some dust blowing from adjoining farms. Ours holding well. FERA men came and started to work at building a new toilet. I regard this work as one of the most important rural projects now in practice. These buildings are constructed according to rules from the State Board of Health. The farmer furnishes the material, new or old. The labor is free.[6]

Sat. June. 8. 35.

FERA men finished today. A very neat job. Treated them to ice cream. They appreciated it. The family to town this afternoon. It is pretty well established that the Blue Eagle is a thing of the past.[7] Not much said about the AAA.

Mon. June. 10. 35.

Among other things that I did today was to plant more sweet corn. I have made frequent plantings of sweet corn this spring in the hope of having a continual supply of roasting ears thru the summer.

Some of our neighbors have finished their first cultivation and are crossing their corn. We have not worked that rapidly. This year we are farming as owner farmers and all of our neighbors are farming as tenants. We must do some things that take time from the fields that their landlords do for them.

Thur. June. 13. 35.

I drove around the neighborhood this morning trying to collect for trucking that I had done several weeks ago, but I did not succeed in taking in any money. This hauling was all done to accommodate neighbors and I did not charge them nearly what it was worth. Their only reason for not paying at the time that I did it and for not paying now was that they did not have the money. Their farming business just cannot be handled in a way that will pay it.

Fri. June. 14. 35.

This afternoon we drove to the county seat. The Loan Secretary has a letter from the Land Bank in Omaha. It seems that there will be additional delays in our loans. Also one of the mortgage companies writes that they will not accept Federal Bonds in payment unless the rate is made higher than the latest Government ruling. More trouble and bother.

City company came today. D.L. crossed the first corn today. Growing things are looking fine.

Mon. July. 15. 35.

If everything goes well this week will be our harvesting week here at Quietdale Farm. We wanted to finish with the corn cultivating today, that is, to get it all over three times. Three cultivatings will be all we will have time for this year and the corn is getting too tall for any more work.

Fri. July. 19. 35.

We finished our harvesting this forenoon. We were in the field in good time this morning and by working steadily all forenoon we finished and came in to a late dinner. It is nice to have this job done. The crop, for us, is much better than some crops we have had, but the oats crop in general will not be as good as some folks have thot it would be.

This afternoon we put things to rights around the place and drove to the county seat. Our Loan seems to be dragging more than I think is necessary. I suppose they are all slow.

Mon. July. 29. 35.

This afternoon I went to town. I was at the court house and the obliging folks gave me my first 1935 corn-hog check, several days ahead of the regular time to pay them out, and I paid delinquent taxes with the money.

Thur. Aug. 8. 35.

I wanted to get some gas tax refund papers made out today and I had to hunt up my permit to do so. We must pay a highway tax on the gasoline we use on the farm and then must have a permit to try and get it back if we can. It would seem that a Nation would be grateful to its citizens that feed it, instead of trying to annoy and persecute them at every opportunity.

Wed. Aug. 14. 35.

The local sealer came this morning and sealed a bin of a thousand bushels of our oats. I took the warehouse papers to the bank and they loaned me twelve cents per bushel on this quantity of oats. $120.00 does not seem very much. Out of this I paid some insurance, the thresher bill and several other smaller ones.

Hardly anyone has horses in condition to haul oats to town. We do not have a permit to truck for hire, but thru this emergency we have been trucking some of the oats in our threshing run on an exchange of work basis and in some cases received a little money on the exchange basis. We hauled only one short day outside of our run and that was where they couldn't get any other truck.

This evening the local constable came to arrest me for truck violations. Before the local Justice, without counsel, I pleaded not guilty. The information was incomplete, not supported by sworn affidavits etc. Later in the evening I found the local attorney and told him to prepare a case for District Court.[8]

Thur. Aug. 15. 35.

I was around the threshing machine part of the forenoon and found an excited angry crowd of farm friends. They were losing money on their oats crop and I was losing money with the truck to get the threshed oats to town for them and then for me to be arrested along with all of the rest of it. Several of them said they would pay any fine that might be assessed against me.

At the village I learned that farmers on the other side of town felt the same way about the affair. A local trucker had prepared the information. Later in the forenoon a state man arrived to take charge of the case. After he had told me all of the things he was going to do about it I suggested that he talk to several of the farmers in the community about the whole situation. He did this. Then I I suggested that he talk with some of the regular truckers other than the one who had complained and he did that too. Then we went to the office of the Justice of the Peace where the State man advised him their charges could not possibly be sustained, that the farmers must be left alone to exchange work, and suggested that we continue as we had been. I did apply for a permit, so that our truck may now be available for any work any place. We never wanted to truck for hire, just use the truck for our own business and right now help thru this emergency. The local attorney was a little disappointed as he was preparing a case to get all of the information into the papers and the courts. A case that would have had a number of truckers and farmers up for testimony and expose the whole farm situation.

Sun. Aug. 25. 35.

Late tonight several cars were prowling around the neighborhood and as I did not have any way of knowing just what they were up to and as they were near our turkey pens, a place they could not possibly have any occasion to be, I went out with the automatic shotgun. Many farmers do this at night and never mention it to anyone. I was well satisfied with the results of my little trip out with the gun and these parties very likely will not bother here for some time anyway.

Mon. Aug. 26. 35.

I went to the county seat this morning and spent most of the day there. The Land Bank Loan and various other matters seem to require much attention. It seems that almost every time they tell us something altogether different about the loan. Perhaps we just do not understand or take too much for granted. I am satisfied that everything will finally work out alright in the end.

Tue. Sept. 3. 35.

The District school was in plain sight from my field and about school time I noticed three cars drive there, taking the children to school. Not any of the three drove more than one half mile and all were cars of tenant farmers and all were shiny and new. Almost a stones throw from the school house one farm family has two children who are physically and mentally very unfit and a half a mile from the school is a home with a child of almost High School age who spends her time in a wheel chair. If the farm income was as it should be these cases would be quite different.

Sat. Oct. 5. 35.

We turn on the radio frequently now to get the war news.[9] There is much speculation as to how the war will affect things in this country, especially as it will disturb things in connection with agriculture.

Sun. Oct. 6. 35.

Along about daylight we were awakened by a commotion among the range turkeys. They are not far from the house and when we rushed out with shotguns we found two strange dogs were among the turkeys and were making every possible effort to kill them. We got a shot at one of the dogs, but did not succeed in killing them. Feathers, dead and injured turkeys and frightened ones were everywhere. Twenty were dead and many were injured. We called in two neighbors to note the loss and condition of things. This took most of the forenoon.

I spent much of the evening working on the farm accounts. I must learn why this business does not pay better than it does, and why all farm businesses do not come nearer paying out than they have been doing.

Thur. Oct. 10. 35.

We drove in to the county seat this morning. We had thot that everything in connection with our Federal Land Loan was working out fairly well. Now it seems that it will require much rearranging on our part to meet their requirements. It will take a little time for us to make these rearrangements, but we can do it. I hope this will be all that will be necessary then. I had understood that the Federal Land Bank arrangements were to assist farmers thru the difficult places, but sometimes I think the mortgage company would have been as kind to us. I am still hopeful that everything will work out right in the end.

Tractor salesmen have been trading the farmers out of their horses until there soon won't be any horses left in the country to speak of.

Fri. Oct. 11. 35.

We are beginning to hear rumors of corn husking and learning more each day about the new corn crop. At this time, it is generally conceded that the crop wll be much better than it was last year, but not nearly up to a normal crop. Many farmers say the corn does not appear to be as near ready for the crib now as it was a week ago. However, a few days of sun and wind can change it very much. I have been thru our fields pretty well and I still think we will have a fair crop. Our work horses are carrying more flesh again and the hogs are making very satisfactory gains since we have plenty of corn for feed again.

Fri. Oct. 18. 35.

All of our family were at the county seat today. We were signing the deeds to the various farms and deeding them around in the closing of Father's Estate. We four boys are assuming the four farms with rather an unsettled state of mind. We do not know just what additional taxes and various added expenses we will have to meet and we do not know anything about how certain our farm incomes will be. We do recall that we are living in a Nation that seemingly has not appreciated their food producers and has sometimes imposed on them to a very great extent. However, since there seem to be some chances for a continued more equitable arrangement for all farm folks we are hopeful that we will make out alright. We know the land and livestock and can do our part. It is the things away from the farm and out of our control that may cause us future difficulties.

Wed. Oct. 23. 35.

Today when I wished to use the party line two farm ladies were discussing the problem of how best to remove the printed letters from feed sacks that they wished to use in some of their sewing work. This is something of a problem for the men too. I heard a farmer remark the other day that the sheets and pillow slips on his bed were made from white feed sacks, as was his night shirt. Where any great quantity of commercial feed is used the bags are sometimes returned to the factory, but on many farms they are made up into various useful things by the ever resourceful farm women.

Sat. Oct. 26. 35.

Today is the corn-hog election day. I think that nearly everyone who can will vote on this proposition today. I handed my ballot to one of the committee men last Monday and I am glad now that I did, for I was almost too busy today to have gone to vote.

Mon. Oct. 28. 35.

The returns on the corn-hog election were the principal topic of discussion wherever farm folks gathered together today. And also among the city folks

CHURCH OF THE BRETHREN, BEAVER, IOWA. 1974.
(L. Edward Purcell Photo)

who fear higher living costs. It has been taken for granted for so long that the farm families will feed the nation regardless, that any arrangement working for a change is viewed with alarm by some city folks.

In our township there were 97 votes for and 17 against. Of the 17 against, ten of them were signers last year and seven had never been with the Adjustment Program. Wherever I go I am still trying to correct the idea that has gotten into the minds of some folks that it is not a reduction program, but an adjustment program. Because of the peculiar circumstances at the time the program went into effect it was used as a reduction program. Possibly in the future it may be necessary to use it as an increasing program.

Tue. Nov. 5. 35.

Jim is having rather an unfortunate experience.[10] He received a small injury in his left hand and infection set in leaving him a badly inflamed hand. He has hired another husker as he will very likely not be able to do any more husking this season. Many farmers suffer various injuries to their hands during husking time. One of the common ones is knife cuts during the early butchering that is usually done about this time of the year. And of course there is always the soreness that comes with corn husking work.

Thur. Nov. 7. 35.

I left the place early this morning and was driving nearly all of the forenoon. I am planning to get a loan on at least a part of our corn crop and I have heard so much about crib requirements this year that I thot I should drive to see the local sealer and learn just what would be necessary.[11] I think I can arrange our cribs without much expense or extra work. In connection with sealing corn I learned that there are only two bins of oats sealed in our township.

I husked a little corn this afternoon. I found the fields entirely thawed out and very wet and muddy. D.L. wasn't in the field at all today. He bot a car today and that took most of the day. He, like many of the farm boys, bot a used car. His turkey sales would justify a new one but he thot he would, as he said, leave the new cars for the town boys.

Sat. Nov. 16. 35.

This afternoon we went to town. Almost everyone was there. The Insurance Company who holds the mortgage has refused to throw off any of the interest. I must change my plans to raise additional funds.

Our County Debt Advisory Committee held their regular meeting today.[12] I sat in with them for a time. They seem to have a very efficient organization. Many farmers have asked me about this committee and my purpose was to learn more about the assistance they might give to burdened farmers. They seem to be preparing for more work in the near future and it seems very probable that they will have it.

Sun. Nov. 17. 35.

This afternoon we drove to town to see a popular movie. Every time I have been in town recently, someone has asked me for money, begging for money for eats, they say. I wonder if I look easy, or prosperous. They all look very needy. So far I have refused them all. They have my sympathy of course. However, many of them will ask a farmer for money, but they refuse to work the hours a farmer must for the pay he receives.

Wed. Nov. 20. 35.

Corn Loan plans are being perfected each day. There is some discussion among farmers as to the advisability of sealing the crop. I intend to seal the greater part of my crop, if the Land Bank will allow.

We drove to the county seat this afternoon and signed more of the Land Bank papers. Gradually things are working to the close of my Father's Estate and the completion of the new land loans. However, nothing seems to be working out as we had originally planned.

Tue. Nov. 26. 35.

I made two new farmer acquaintances in the city this afternoon. Like myself, both of them are becoming farm owners in the same way that I am. All of us agree that the greatest difficulty we may have in farm ownership will be high taxes.

Corn husking is pretty well all done. Many report a shorter crop than was expected and a lower quality also. Next season's seed crop is an increasing problem.

Thur. Nov. 28. 35.

With Thanksgiving Day past and the year drawing to a close, I am trying to plan some way to save as much of the grain and live stock that we have left and get the old mortgage out of the way and then plan some way to get thru next year. Under ordinary conditions, after that, we should get along fairly well.

Tue. Dec. 3. 35.

We finished the husking today, this afternoon. Bill helped part of this forenoon. The help went home early this evening. Our crop was not as large as I had hoped it would be. The quality was better than on some farms tho.

This evening I am busy planning a way out of the mortgage and loan difficulties. I have always tried to make it a point to help everyone wherever I could and especially farmers, but I cannot think of any place now where I could go for much help if I should find myself in a tight place. There will be quite a lot of satisfaction in getting thru alone if I can. Tomorrow I will begin on the job.

Wed. Dec. 4. 35.

I went to town this morning to see the bankers. It is becoming apparent that I must sacrifice grain and live stock to satisfy the mortgage holder, so that I can get the new Federal Loan thru. I wanted to learn if I could borrow money at the bank for new stock a little later, but I did not get any satisfaction.

At the county seat I learned that our farm was sold for taxes Monday. Tomorrow I will arrange a corn loan and redeem the taxes. But I deplore a system where the corn loan and tax sale dates are arranged this way. Our taxes were just one year behind, not two or three. The tax sale lasted a day and a half.

This afternoon I drove to another town to advertise some pieces of farm machinery for sale.

The sealer will come tomorrow to seal the corn.

Thur. Dec. 5. 35.

Today was a dark, cloudy day. I leveled the corn in the cribs and the sealer came about the middle of the forenoon. He had finished his work by noon and was with us for dinner. He is a farmer, an elderly man, who lives in the opposite corner of our township. It seems to me that the method of measuring the crib is not entirely fair to the several farmers.

This afternoon we went to town and cashed the loan papers and deposited the money in the bank. Then I went to the court house and paid the taxes and extra expense and thereby redeemed the farm as it was sold for taxes Monday of this week. The corn loan was arranged several days later than the tax sale date. Sometimes I think there is more blundering than efficiency among some of our organizations. The Farm Bureau should have been looking after this. Things of this kind is one of the things we pay a membership fee for.

The loan secretary is going to write the mortgage company in an attempt to get our interest reduced. He was not very optimistic about this, but I am hopeful.

Fri. Dec. 6. 35.

Some rain had fallen last night and the weather today was quite damp and misty. I did odd jobs around the farm and attempted to plan some way to raise additional money to apply on the mortgage.

This afternoon I went to town for help and advice and came home this evening feeling a little better. I know that many other farmers are in worse shape than we are and that many more never will be able to get things fixed up and keep their farms unless something different is planned soon.

Business men complained of very unsatisfactory conditions. I think this is to be expected in a farming community. Low prices on what the farmer has to sell and high ones on the things he should have but does not have to sell must certainly create this condition. Then too many farmers are beginning to plan to operate on a very economical system and this is temporarily bad for the merchants.

Tue. Dec. 10. 35.

I drove to the county seat this morning in an effort to find some way to more easily meet the demands of the mortgage company and to confer with the loan secretary.

This afternoon I attended a farm sale conducted by one of the near neighbors. The sale was well attended and things sold fairly well. I clerked nearly all of this sale. The first one that I ever clerked. It was interesting work. I have studied farm and community sales so long that this gave me an opportunity to try out some of the ideas I have been turning over in my mind for some time.

Between times today and this evening I have been rather depressed with our own situation here at the place. I am still hopeful of finding a way out that won't leave us in too bad a shape.

Wed. Dec. 11. 35.

This morning I sat down and wrote a letter to the mortgage company asking for a reduction of a small amount on the interest. The very least I would need was all I asked for. However I do not think they will grant it and later in the day I drove to see a neighbor to try to sell him some of our live stock and machinery. The things that we can spare the easiest. I will not have a farm sale. I do not think it is necessary to go that far and we must have something left to farm with next year so that we can make the loan payments. It is a very unsatisfactory arrangement all around.

A dairy report came from Washington this afternoon in the mail. In it I notice that the consumption of Oleo has increased 72% this year. I like to think of what cows would be worth if butter consumption increased that much. Also I am much interested in a set of figures showing the tremendous increase of the imports of farm products this year. If the figures I have seen are true the farmers surely have been played for a bunch of suckers.

Thur. Dec. 12. 35.

After I had finished the morning work I took a good look at our drove of hogs and then stiffened myself for the job I had to do. I drove to a neighbors and told him just how I was fixed, so he came and looked at our hogs and we finally made a deal that is the best I could hope for. We loaded the hogs and took them to his farm. I try not to think of the loss this apparently is going to be to us, but my mind keeps turning over the figure these hogs would have brot if I could have fed them out to maturity. And the effect this amount of money would have had on our future farm operations.

This afternoon I went to town to the bank and to the loan secretary's office and am trying to make the money reach as far as it will.

I am still confronted with the additional amount I must raise yet and trying to think of some place I can get it. There is still some oats that I should keep for feed and seed that I can sell for eighteen cents per bu. But there is a twelve cent loan on them. Then there are the milk cows and some of the other feed.

The low prices of several years ago, the drought last year and the low prices of some things and the scarcity of others are among the things that are pinching me so hard and will greatly hamper many other farmers.

Fri. Dec. 13. 35.

The mail today brot a letter from the mortgage company saying that they could not make any reduction in the interest. I regret this very much. I am going to feel this pinch considerably. Also a letter came from the Federal Land Bank suggesting that I close my loan. This afternoon I drove to town to make more plans to do this.

When I had finished the business as far as I could go with it at this time, I went to the community sale barn for a short time. The sale was in progress, but not quite as good as some of the former sales.

Mon. Dec. 16. 35.

I spent the morning going over the mortgage and loan figures and had a plan worked out that seemed as if I could get along fairly well. But I was uneasy about it as we drove to town. In town I found that if things had not been deliberately misrepresented to me I at least had not been fully and correctly informed and the amount of money that I now must raise is almost more than I can possibly get together. I spent a long time with business advisers and came home much discouraged. I had expected something like this but I did not think it would be as harsh and rough as it is. This evening I am rather bitter toward the local, State and National agencies that are supposed to assist the farmer. I know something about how thousands and thousands of farmers have been treated and how they feel. Tomorrow will be another day. Perhaps I can yet work something out to get by with.

Our county Farm Bureau held their annual meeting today. It was not well attended. The poorest meeting in years. The organization has been directed so long by someone away from the farm that I fear it is slipping.

Wed. Dec. 18. 35.

This afternoon I drove to Perry[13] to see the offices of the Production Credit Corporation. They disregard any equity one may have in land. They make their loans on the number of head of cattle, sheep and hogs, and the value of them, that the borrower may have, loaning 75% of the appraised value of these stock and require the mortgage to cover this stock and also the horses and machinery of the borrower. They stated they were supposed to assist only needy borrowers but admitted they frequently failed in this. I had thot they might assist me in my refinancing when my loan is finally completed. Their interest rate of 5% and their bank stock of the same amount makes a rather high rate of interest. However, many farmers could use their funds and make money by so doing. I looked in at a community sale a short time this afternoon. The character and progress of this sale seems to indicate that hard times are still with some folks. I think small town people rather than farmers were most active in this sale.

Sat. Dec. 21. 35.

In figuring a way out of my refinancing I had decided it would be necessary for me to sell the sealed oats and this morning I drove to the village to see the manager of the Quaker Oats elevator there. He was paying twenty and twenty-one cents per bushel for oats. When we unloaded the first load they were found to test 27 pounds per bushel. We continued to haul until enough were sold to pay the sealing note, receiving twenty cents per bushel for those we sold. After the required amount were sold we checked up on those remaining in the granary bins and found that the sealer had not made any error in the measurement of the bin, but the crop as he sealed it would not weigh out the measured number of bushels. We have enough of the crop left for next spring seeding and some for feed. We received twelve cents per bushel from the sealing and sold them at twenty cents. Perhaps a cent or two less than the market price at threshing time. The sealer received five dollars for sealing them. There was a recording fee of fifty cents and the interest amounted to two dollars and forty-five cents. Anyway, by sealing them we had the money to use through the emergency. This afternoon I went to see the bank and pay the oats loan note. There aren't any new developments in the land loan.

**CLEARING A COUNTRY ROAD.
WINTER OF 1936.**
(Courtesy of the
Iowa State Historical Department/
Division of the
State Historical Society)

CHAPTER 4. **DRIFTS**

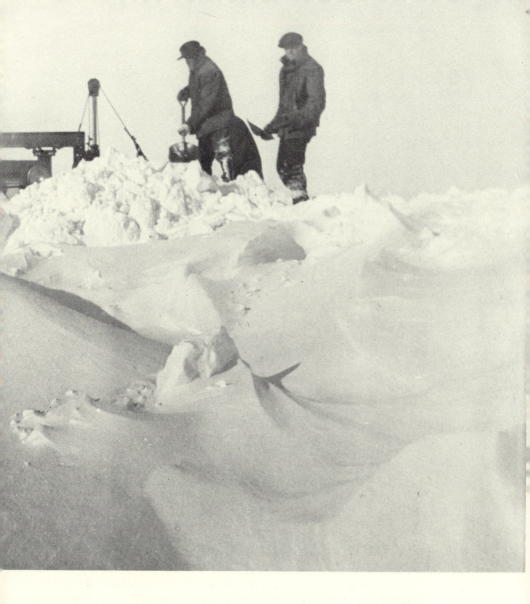

DEBTS, AND DROUGHT: 1936

Wed. Jan. 1. 36.

Today is the beginning of another New Year. And we farm folks are beginning the year with thots that are a mixture of hopeful anticipation and uneasy fear. Experience has taught us that some of the years are very unkind. The land and the farms usually are very good to us. Sometimes the weather and the markets and our other business associates are not fair or even decent in their treatment of us.

Our increasing taxes are one of the many things that worry me. For the year of 1935 we have paid in excess of $2.00 per acre for the various taxes of all kinds. This is too much for a 160 acre farm. And if taxes worry me I know they must worry many other farmers.

Thur. Jan. 2. 36.

In town I learned that the old mortgage company will not make any reduction in either interest or principal payments. This is, I believe, final and I do not expect to have any further communications with them in the matter. I do not know just yet what arrangement I can or will be able to make now, but think I can work a way out.

Fri. Jan. 3. 36.

Quite a busy day today, made of odd jobs around our place, helping Bill doctor a sick horse, driving to the county seat, attending a community sale etc., and tonight listening to the President's Message To Congress.

One of Bill's horses is getting rather old, in fact many of the horses are becoming quite old, and for some time Bill has had quite a time to get the horse up after it has been down several hours. Sometimes in the field and sometimes in the barn. But this time it was really sick and I told Bill so. However, when he did finally get the vet it was too late. This leaves Bill with only two horses. Many farms that formerly had seven to nine horses have only three or four now.

Sat. Jan. 4. 36.

We went to town tonight. It seems quite some time since our family went to town on Saturday night. The weather today had been an average winter day, but tonight it seemed very cold again. I looked at the thermometer when we were back home but it wasn't down to zero.

I saw and heard Secretary Wallace in the news pictures at the movie show. It seems to be generally conceded that the United States Supreme Court will announce an opinion in the next few days that will have a definite bearing on the AAA.

Sun. Jan. 5. 36.

Winter has settled down on our community. A little more snow fell today. The whole countryside is entirely white. Here and there farmers are beginning to use their bobsleds around the farms. There would be good sleighing if there were any need for it. Snow plows keep the main roads clear and auto travel continues uninterrupted. Farm marketing is being done more with wagons again. Truck laws and the economic situation have almost done away with farm trucks. This works quite a severe hardship on some of the farm horses.

Mon. Jan. 6. 36.

Right now many people look at today as one of the dark days for agriculture. The Supreme Court Decision came out today on the AAA.[1] Time will tell whether or not it was a dark day. But this much is certain, things are in pretty much of a mixed up state of affairs just now. In a few days or a week things will begin to reshape themselves again. Just how and in what way seems to be rather uncertain tonight.

We had another winter day again and here at the place we sold the baler to the party who looked at it running when we were baling last Saturday. In some ways I regret to see it go. The money I received for it will be very useful just now in closing up the loan affairs.

We were in town this afternoon for a short time and stopped by a coal mine. It is a very busy place just now. There is some scattered fuel on almost every farm, but getting it out of the snow is disagreeable work.

Tue. Jan. 7. 36.

Ten above zero this morning. Some wind blowing. Winter has a good grip on the community. Very little activity. No outside work today. Just in the barns this forenoon.

I attended a community sale this afternoon. A small offering, sold in the average way. Quite a large attendance of farmers, but they were more interested in the AAA than the sale and many small groups were exchanging opinions quite freely. There was much complaint about how the corn-hog program had been handled locally. Farmers have hesitated to talk about the local adjustment and allotments, because they thot it might affect their checks. Now they are free to express their opinions. All of them seemed to agree that the program had been useful and did them much good and was worth while even if it was not allowed to go farther.

There is every indication that the winter weather will continue.

Wed. Jan. 8. 36.

The wind was blowing from the southeast this morning and a thick heavy snow was falling. I drove to town to settle old accounts with several business men. Two of them were very nice about closing up the old business and planning for the new. But one of them seems to have a different outlook on the future. I am very glad tho that two out of the three can look at things as they do. I feel that their attitude regarding me is typical of their attitude regarding all of agriculture.

By noon the snow storm was over and I trucked a load of baled straw to a sale where I refused the bid of 15 cents per bale and finally sold it by the ton privately.

Only poultry and various articles made up this sale offering, which was a large one. Also the crowd of buyers and spectators was a large and interesting one. Many were there to meet old acquaintances from near and far and to exchange opinions on farming, the outcome of the AAA and politics. All of them agree that the AAA has served a very useful purpose and many of them hope it will be continued in some form or another. All hope that the balance of the '35 checks will come out soon because all of them had planned some specific use for them. Business men are beginning to be afraid conditions will become as they were before the corn-hog checks began coming to the farmers.

Thur. Jan. 9. 36.

Township corn-hog committee men that I talked with seemed to be closing up the business. The county agent was not so sure there would ever be any more checks. Here and there a farmer says he is glad it is over with. Farmers as a rule do not seem to be condemning the Supreme Court. In a way I do not think they are familiar with it. They do accept the decision as they would unfavorable weather or something of that nature.

Fri. Jan. 10. 36.

Was at the new sale barn a few minutes today. A big crowd and values still holding up well, tho a cautious note was running thru all transactions. Here and there I hear a farmer saying that he has lots of feed on hand.

At this sale crowd I thot opinion was changing and growing for a continuation of the AAA or a similar plan. Business men are of an uncertain mind and farmers remind them that there isn't a surplus staring us in the face anyway.[2]

Sat. Jan. 11. 36.

Numerous rumors are around about the future of all things connected with agriculture. Several local prominent business men said today that business was dead. At any rate the county farm debt committee had a hectic day. One neighbor who had an application with the Federal Land Bank has a letter today that his application is refused. The local grain elevators are shelling their small stocks of ear corn. It is reported that today is the last day for making corn loans and sealing corn. Distant landlords are rushing to get their sealing done. Other talk is that loans will be made at 35 cents per bu. There is talk that existing loans are cancelled. I think banks who were very anxious to get corn loan paper several weeks ago are shying away from it now. Several more days like yesterday afternoon and today and things locally, and I suppose everywhere, were in a very tangled and difficult state. A few more interesting days are ahead of us.

Tue. Jan. 14. 36.

I think I finished the loan business today. Among various things I receipted for the Federal Land Bank money and the Secretary sent it to pay off the old mortgage. After a year of work, preparation and planning the job finally seems to be done. All of the business people connected with it have been very nice and accommodating. I do not know just what comment to record in this connection so will pass it without any.

Here at the place this evening I hardly knew what to do with myself. Tomorrow I think I can begin to plan for a new future. I expect plenty of trouble and will try to prepare for it, but I am hopeful that future affairs will come along in good shape. At least my troubles will be the same as thousands of others. However, there isn't much satisfaction in that.

Fri. Jan. 17. 36.

This morning all weather signs seemed to indicate coming bad weather. We scraped our roads again, clearing them of the accumulated snowfall. Then we drove to the village and bot a small quantity of coal. Ordinarily we burn only wood, but thot it better to have some coal in case our wood supply would become exhausted.

By noon only a light snow was falling and we decided to drive to the county seat to see if anything had been overlooked in the loan business. The drive to town was without incident. In town I found that so far all of the business had been completed. While we were in town the snowstorm broke and we came home thru a blizzard. The worst one of the season so far. After we were out of town we could drive without any trouble.

Here at the place we made the stock comfortable and things ready for a heavy snow and anticipated very cold weather to follow.

From the large collection of weather calendars and almanacs that I have accumulated I would say that we will not have a very good farming season this coming year. That is, from the weather standpoint. Too much cloudy weather and too much moisture is indicated.

Sat. Jan. 18. 36.

A typical old fashioned winter day today. The sky was clear and the air cold. The thermometer registered twelve below zero this morning and did not get up to zero all day. It was a winter day as we used to have them a long time ago. The kind of a day that my Father and my grandfathers would have appreciated very much. Good sleighing weather, a good time to do the hauling and a good day to butcher beeves or for hog killing. It seemed to me that they always selected a day like this for the work I was most interested in watching them do.

The blizzard of yesterday and last night left many roads blocked with snow and it will be a day or two until the big snow plows can get over all of the highways. We are fortunate that our roads did not close. I am hopeful for warmer and better weather next week. There are so many things that I would like to do around the place when the weather will permit. On the radio today I heard that Secretary Wallace will have a new corn-hog plan ready for adoption by February 15th. This will be quite satisfactory to many farm folks.

Sun. Jan. 19. 36.

A very cold day today. Fifteen below zero this morning. By noon the thermometer stood at zero and at chore time this evening ten above. We were indoors all day excepting when taking care of the stock. It is hard work to care for livestock and poultry when the weather is so cold.

Community activities are almost at a standstill. Our farm ladies attended church services in the village church this evening and reported an attendance of twenty-six persons.

The snow plow, working on nearby roads today, got as far as Jim's place and found the drifts too deep for it to handle. Several years ago when the snow was deep, plows were in operation day and night until the roads were clear, but this year they are not working so steadily.

While I was at home alone this evening I happened to look out of the window and saw a dark red light on the sky. I knew it would mean only one thing, a farm fire and very likely a farm home. Using the phone I learned that a farm house was being destroyed. The ladies returned from church about that time and Bill and I drove as near the farm as we could get and walked the remaining distance. The snow plow had not been on nearby roads. The fire had started from a defective flue and the entire home destroyed. A few things were saved from the lower story, but the upper story and basement things were lost. Fortunately the wind was blowing away from the other farm buildings and they were not lost. Only a few neighbors were aware of the fire or able to get to the unfortunate farm.

Wed. Jan. 22. 36.

Today will go down in history as one of the coldest days on record for many years. The thermometer was variously reported as being around twenty-five to thirty degrees below zero this morning and the highest it recorded today was around eighteen to twenty below.

School attendance was as low as the thermometer. Many schools were closed. All traffic suspended locally and much of the thru traffic tied up. No mail anywhere today.

The bitter cold was very bad for all livestock and especially so for that in transit. Our stock suffered some too and they consumed enormous quantities of feed also.

The cold wind sweeping down from the north chilled everything and everybody. The sun made a feeble effort to shine but there was so much cold and light snow in the air that almost no warmth reached the earth and tonight there are indications of as low or even lower thermometer readings than there were last night. The coldest day in twenty-five years and forty below predicted for tonight. The severe weather is causing much suffering and a great economic loss.

When we were not working to make the livestock as comfortable as we could, we did carpenter work in the house, changing a partition. As we can do this with an expense of about a dollar for new material we thot best to do it. Otherwise, if more money would have been needed we would have waited until some other time.

Thur. Jan. 23. 36.

Another very bitter cold day today. Thermometer below zero all day, has been for several days. From 18 to 25 below according to various thermometers. Community activities remain pretty tied up.

Fri. Jan. 24. 36.

Another of the below zero days. The sun was shining much of the time, but the air was cold all day. Cattle could get a little warmer along side of the barn where the sun was shining and the wind did not blow.

The snow plow, which was stuck yesterday the other side of town, came into our neighborhood today. The machine they are using now is much too small for the work it has to do. Several years ago a much larger machine was in use and it could travel thru any drifts it encountered. The present one cannot do this.

Sat. Jan. 25. 36.

Still another cold day. Below zero again today, all day and this evening.

Just out of our community a farm home was left without a mother today. There are four children from one to fourteen years of age. Three boys and one girl. The flu, the severe weather, no phone and the bad roads were a combination that took a very much needed farm wife and mother. This is one of the saddest farm tragedies that has come to my notice.

Sun. Jan. 26. 36.

Today was a quiet Sunday on the farm. The phone and the radio was our contact with the outside world. For about an hour today the mercury climbed above the zero mark then started back down again.

Our ground feed has all been fed and until the weather becomes warmer we must continue to feed the livestock and poultry whole grain. This is a common practice on many farms but on the whole I do not think it is a profitable one.

If the cold weather continues next week I do not know how to plan the work for accomplishment. The cold weather has already caused much delay and extra expense for all farm folks and if it should continue until spring it would work a real hardship on some.

Another farm home burned yesterday forenoon. There are more than the average number of farm fires the past few weeks. I am rather inclined to blame the house fires on the women folks and farm barn fires on the men. Just a little more care and attention and these things could be avoided in many cases.

Sun. Feb. 2. 36.

Today is "Ground Hog Day" and I would say that the ground hog was not able to see his shadow today. I do not think farm folks pay as much attention to the weather signs now as they did in former years.

As our roads are still full of drifted snow we could not get away to visit or attend services and spent the day at home. The thermometer climbed to several degrees above zero at one time, but tonight it is working back down to one of the lower levels.

Our radio quit working yesterday and we as many other farm families are pretty well cut off from the outside world.

Mon. Feb. 3. 36.

The bad weather continued today, in fact it became much worse. A northeast wind was blowing and the thermometer was below zero this morning. Later in the forenoon the wind shifted to the north and light snow began to fall. This snow began to fall faster and heavier and by dinner time we were having the ugliest blizzard of the season.

We kept all of the livestock in the barn and tried to water them there, but several head refused to drink water from pails and we gave up trying to get them to drink.

D.L. was worried about the large flock of turkey hens. He knew that whatever happened to one of them would very likely happen to all of them. Meaning the loss of one or two in the storm would mean the loss of the flock. However, at roosting time he found them all going up in the large trees and they finally all settled for the night in an out of the way place in the grove and he was satisfied that they would wear out the blizzard safely.

Living is becoming rather a problem for many farm folks. As the years have gone by more and more farm people have been depending on the stores in town for their living or a greater part of it at least. Now with blocked roads the groceries etc. soon run low. Fuel is a problem for some and feed for the stock for others. The days continue to go by without any warm, thawy weather. Apparently there isn't to be any let up of the cold weather until spring comes.

Tue. Feb. 4. 36.

For the past several nights I have been sleeping in a reclining chair until midnight, then going to the barn to look after the stock, but last night I did not go to the barn in the night, and this morning I found a new calf. As the thermometer registered 20 below zero I took the calf to the basement and will keep it there until we have warmer weather. Our barn is not warm enough for these new arrivals this winter.

The blizzard had worn itself out by noon and this afternoon I butchered a young beef. Wearing a pair of the new rubber gloves that we buy at the newer stores, gloves that are both warm and waterproof, I found the work quite agreeable in this severely cold weather.

If yesterday's and last night's blizzard will be the last one of the season everyone will be well pleased. Spring is coming closer each day and very little of the seasonable work is being accomplished.

Wed. Feb. 5. 36.

Very cold again this morning, 24 below zero. I need a hair cut and a shave quite badly. Our road is completely filled with snow. This afternoon ten neighbors got together and scooped out three miles of it. It is surprising how much ten men with scoop shovels can do at moving snow. These drifts were five feet deep in places. Cars can get thru this evening. It requires careful driving tho. All of the men shoveling snow needed shaves. Many of them will attempt to get to town tomorrow. The County Supervisor promised to have a snow plow in the community tomorrow. If so we have enough work done that it can get thru without any trouble. The same ten men have agreed to accompany the snow plow thru this road whenever it can get here. Driving to Bill's with the bobsled I picked up our three days mail and tonight we have plenty of reading matter again.

Thur. Feb. 6. 36.

20 below again this morning. Some nearby farms reporting a lower temperature than this. Our stock fairly comfortable regardless of the very cold weather.

The snow plow arrived and worked on our road this afternoon. All of the men living on this road were out to accompany and assist if needed. We all feel better and safer tonight as we are again connected with outside assistance in case of accidents, illness or farm fires.

Sat. Feb. 8. 36.

Yesterday afternoon and last evening the weather was quite good compared to what we had been having, but some time in the night last night the wind turned to the northwest and this morning we were having what turned out to be the worst blizzard of the season and one of the worst for many years.

The wind blew harder, more snow fell and the drifts piled higher and the cold was more intense than any storm that I can remember of. We thot that the storm would quiet by evening but it did not and in fact became worse, the wind blowing harder.

We did what we could to make things as comfortable as possible for the livestock and the poultry, driving the flock of turkey hens into the big barn and giving everything extra feed and bedding.

Our roads that we had worked so hard to clean are drifted more badly than ever. Travel in the highway has ceased and trains are running with much difficulty. Late tonight there are indications that the storm will continue tomorrow.

Sun. Feb. 9. 36.

Stormy weather all day today, and we were at home all day too. After I had finished work with the stock I spent my spare time reading and looking up weather statistics. Going back over the history of this particular farm I find there have been many unusual instances and many variations. I am the third generation of the same family that has been here during the past sixty years and I can recall and recollect the things my Father and also my Grandfather talked about in connection with the weather. Droughts, floods, unseasonably warm and unusually cold and numerous unequalities have followed one another with a regularity that we soon forget until we begin looking into the past.

With so many of the highways and the roads closed with snow each farm is almost compelled to look after itself. First of all we are all very careful about fires, then accidents and lastly our water supply for the livestock. Our stoves, lights and wiring are attended to very carefully because if a fire should start it would mean disaster. Then we think, move and act with particular care to avoid accidents. We could not get to a Doctor and he could not get to us, so we are careful all of the time. Lastly we watch the pump very carefully. Our well is a four inch casing and if the pump breaks we could not draw water with a bucket and could not get to town for repairs.

Mon. Feb. 10. 36.

Very cold weather again this morning. Way below zero again. We didn't try to do anything but care for the stock this forenoon. This afternoon we scooped snow on the road, working toward a neighbor's place, where other men were scooping to get to the highway. When we quit this evening to do the evening work we had cleared enough road that by noon tomorrow I think we can get away with the car. We drove a team on the road today, but tomorrow we will take the bobsled. Just before evening I drove over to Jim's place and got our mail. He got it from someone's box and phoned us.

We must get to town tomorrow. Two weeks ago something went wrong with the light plant. Several days later the high test gas was all [gone] for the lantern, then the kerosene was all and we used one kerosene lamp. The same way with the groceries. Flour, sugar, coffee, tea, soap, matches etc., were nearly all or entirely used up and we kept eating something else from the canned meat and fruit from the cellar. And many of our neighbors are not getting on as well as we are.

Tue. Feb. 11. 36.

Cold weather again today. We drove the car to where we quit scooping on the road last evening and began working again and by noontime we met a bunch of neighbors coming from the other way and our roads are open again. Every day I phone to the county supervisor in charge of our part of the county about the progress the snow plows are making, and he tells of the broken and disabled machinery and the tired crews of men who are working with them.

We drove to town this afternoon and returned late this evening. We drove in a round-about way to get to town. We had broken and loaned all of our scoop shovels and drove to town without any way to scoop out with. In town I found that the county had bot all of the shovels, but three had come from a wholesale house at noon and I bot one of them. At the grocery store I bot a large supply of groceries and had them charged. The first time I did not pay cash for years and years. Business is at a standstill. Crews of men and tractors working to keep mine roads open. Little dabs of coal trickle out to towns and families. No school anyplace. Doctors getting to patients in bobsleds and walking. No favorable weather in sight. Altogether one of the worst situations we have ever had. We have another new calf and I carried it to the basement right away.

Wed. Feb. 12. 36.

Another of those discouraging winter days that we have been having so many of this winter. There is more snow and deeper snow now than our oldest residents can remember of. Everyone has been hoping and praying for days for a break in the severely cold weather and a let-up of the snow fall, but it does not come. Life is a problem of existence until spring. Groceries and food for the people and fuel to keep them warm, then feed for the live stock, which is gradually getting weaker and thinner. Many farmers who must move this spring are selling their cattle. Others who have sufficient feed but cannot get it to the stock are selling them. Our hogs have been penned in the hog barn for weeks because they would walk on the drifts over the fences if they were outside.

Thur. Feb. 13. 36.

This afternoon I cut more wood and hauled a load of litter to the fields and walked and drove across the fields to Bill's place to get the accumulated three days mail. There wasn't anything pleasant or cheerful about this trip. The cold was intense. The roads were entirely blocked and the horses toiled thru the deep snow as I drove them across plowed fields and meadows, then I tied them around to the sled and walked the remainder of the distance to Bill's. Our countryside is a bleak wilderness. As desolate as the frozen northland. And we cannot find any encouragement that warmer weather is near at hand.

Sat. Feb. 15. 36.

Today will be a day that many of us will remember for a long time. The snow plow came over a part of our roads today. We were on the road all day. We made every effort to keep it traveling as much as possible and all of the men on our three miles of road were out to meet the plow and shovel out the bad places ahead of the machine. We did keep it traveling steadily only [except] when we stopped for coffee one time and for lunch another time. At one place in the road where drifts were very deep we took the machine thru the fields and this brot it close to our place, where we stopped for the first time for hot coffee. Then we continued and by noon were at Jack's place where a short stop was made for dinner. Then on again where we met fifteen men who were to work with our ten, but these men seemed to rather stand and watch the machine than work to help it. We finally turned on a road where the snow was very deep and as a snow storm was coming up the outfit was turned around and taken back to town. The eight families on our three and a half miles of road consist of exactly 40 people. Our men worked like veterans and were always ahead of the machine scooping the deepest snow. They continued on past their own farms. I am of the opinion that if the men farther on the road would have been more inclined to assist, the plow crew would have continued on farther.

Sun. Feb. 16. 36.

Way below zero again today. Yesterday was a very cold day also. Other than the necessary work we rested today. Everybody was tired out from yesterday. Pausing in my shoveling yesterday I looked around at the men and counted ten of them at one time who had white frozen spots on their ears, or noses or cheeks.

Last night when the evening chores were finished we drove the car to the village for groceries and then went to a larger town for a radio battery, there, after I had seen these roads, I bot another additional supply of groceries. The transcontinental highway that we drove over for a short distance last night was hardly open at all.[3]

Unfavorable weather conditions tonight indicate that we may lose the road we have worked so hard to create.

Mon. Feb. 17. 36.

A very cold and stormy day all day today. The thermometer was 16 below this morning. A fairly heavy snow was falling and a strong wind was blowing, drifting both the old and the new snow.

Wed. Feb. 19. 36.

The weather was a little better today. Not quite so cold this morning and the sun shone some warming the air slightly. Jim came over and we hitched to our road grader and attempted to clean the snow from the road. The grader was not entirely suitable for this work, but we did get more than a third of our three miles open with it. Jack and the boys are working with shovels on their end of it and we shoveled this afternoon. Bill and Ernie were shoveling too. Sometime tomorrow we will have a car road again. Some way we all agree that perhaps this will be the last time we will need to "scoop out." We hope for more favorable weather from now on.

Thur. Feb. 20. 36.

Fair weather today. Clear skies and slightly warmer. One of the most persistent worries of the stockman is an abundant supply of drinking water for the livestock in the winter time. We have been very fortunate in this respect until today. Our pump quit working and when we assembled tools and lifted it part way out of the well we were able to repair it and get it in commission again.

The remainder of the day we shoveled snow in the road. By chore time this evening we had the road clear again.

Sun. Feb. 23. 36.

The long hoped for and looked for break in the winter weather came today. The sun shone all day and the air seemed quite warm compared to what we were accustomed to. The snow melted and settled down but I did not see any place that water was running.

The school bus drove over the road this afternoon, just to see if it would be possible to resume school again tomorrow. The bus made the trip thru in good shape and plans are under way to open the village school again.

All of the farm stock enjoyed the warm weather very much and it will take much of it to get them back into proper condition again.

Tue. Feb. 25. 36.

Our road has been open for cars for several days, but there are many roads in the community still blocked with drifted snow and a new snow plow came in this morning and did a little work near us so that the school bus would have a better road, then went on to clear other roads.

This new plow is a four wheel drive truck with a plow on it. I learned that it cost the county $6,500.00 and the plow is extra. The truck gives it a faster traveling speed than the crawler type of tractors and it does very well plowing in slow speed. We had one stretch of 50 rods of snow, five feet deep, with a thick frozen crust on the top of it and the outfit nosed its way thru this 50 rods in 40 minutes, then galloped on to another place. The old plow we used last week is said to be wrecked beyond repair. Now the community is preparing to resume normal activities again.

Wed. Feb. 26. 36.

Sometime between midnight and morning last night several inches of a wet, heavy, snow fell. It was too heavy to blow and drift and we did not mind a little more snow, as long as it would not pile up. But before noon the wind changed to the northwest and by midafternoon we were having another blizzard, which turned out to be about the worst one of the season. We had hoped that the wind would quiet at sundown, but it did not and there is every indication that all roads will be closed before tomorrow morning.

Community life will be at a standstill again. Farm sales must be delayed further, schools discontinued etc. I do not look for a long cold spell this time. The greatest inconvenience this time will be the delays about moving time.

Thur. Feb. 27. 36.

The wind had stopped blowing this morning and we are going about the business of working ourselves out of the results of another blizzard. The thermometer did not get down to zero this time and by midafternoon the sun came out but it did not thaw very much.

I worked around the place this forenoon and found drifts as high or higher than any we have had yet. I phoned to the highway tool shed and learned that all roads are drifted quite badly again and that it will be a long time before they can all be plowed out. The plows will attempt to get to farms where sickness and moving need first attention.

Mon. Mar. 2. 36.

The weather was much warmer again today. Some indications of a general thaw. We worked around the place cutting wood, etc. Also I worked on the roads scooping snow, helping a neighbor in a large drift, so that when the snow plow does get here it will be sure to get thru. Learned this morning that the snow plow worked alone last night. Farmers in the community where it passed thru did not give any assistance.

Wed. Mar. 4. 36.

This evening we learned that the big snow plow had been repaired and was out of town for a short trial run, then it was headed our way. I phoned all of the farmers on our road and a few of the first of us met it. As we progressed down the road we continued to pick up men until we had 21 of them. Sometimes we rode on the back of the truck and when necessary worked ahead of it, breaking out the heavy crust or shoveling the deeper drifts. Then when we were thru and to a clear road we all loaded on and traveled in a fast speed to the next drift. Around midnight we turned the plow around and began working back

nearer home, dropping off farmers as we passed their place. At one thirty
o'clock we came to Jim's farm. There was only Jim, Hank, D.L. and myself and the crew of two men left and we went in to Jim's for lunch and
coffee. We almost had to lift the driver down from the cab. He had been
driving for more than ten hours.

Thur. Mar. 5. 36.

After lunch we left Jim's place and continued on, clearing the remaining
distance on Jim's road. Then D.L. and I came home and went to bed at three
o'clock this morning. Jim and Hank went home and the plow continued on
alone. When I had finished the chores at eight o'clock and phoned the county
shed to report our night's work, the plow wasn't in yet at that time.

We will all remember last night. Early in the evening all of the men were in
good spirits and jolly. Every farm house was lighted and our progress closely
watched by the farm ladies and the children. However, along toward midnight,
after we had toiled thru many deep drifts and the men and plow crew were very
tired and the weather had turned very cold it wasn't nearly so pleasant. Many
of the men did not have their heavy coats along and some of them became
chilled. Several times we came very near to having a bad accident and
someone seriously injured. At one place we took the plow over a long stretch of
flooded roadway. The plow was one of the new modern four wheel drive truck
outfits and worked very efficiently. When we men rode in the truck box we
were with a load of barrels and miscellaneous tools and gear and after midnight
this seemed a very uncomfortable place to ride. Working in the very deep drifts
seemed better than riding in the load.

This afternoon we were all rested and many of us went to town and began to
take up our community and private affairs again. I saw the first robin of the
season today.

Sat. Mar. 7. 36.

The weather was very nice today and we drove to town this morning. A free
movie and a free lunch this noon, all by the manufacturer of a new all-crop
grain harvester, were the principal attractions of the day. We did not have time
to see the pictures and missed the lunch. Many farm folks did enjoy both of
these features tho.

We drove our big truck to town today and traded it for a smaller one. Our
large truck was almost too large for ordinary farm use and our trucking laws
were not planned to assist the average farmer, rather they were planned to
annoy him. Our immediate neighbors will be the real losers because they will
lose the service of our truck at a price they could afford to pay. The smaller
truck will require some minor repairs quite soon.

The brilliant sunshine made short work of all but the larger drifts today.

Sun. Mar. 8. 36.

The radio today brings news of the changing European threatened war conditions.⁴ A serious situation seems to be developing quite rapidly. Just what, if any, effect it will have on our farm conditions remains to be seen.

Thur. Mar. 19. 36.

We drove to the county seat this forenoon. I bot another used tire for the truck. And I sold some old brass. There seems to be more activity in the junk business than there has been for some time; war possibilities stimulating the metal market I suppose. I canceled my hail insurance policy today. Perhaps I should not have done this, but the farm income can be made to cover only so much.

Tue. Mar. 24. 36.

Bill and I drove over into another county this morning to look at some farm machinery Bill was interested in. While we were there we learned several things about recent developments affecting the farm situation. The grain market has declined again and 16 cents per bushel was the elevator price on oats. This caused much uneasiness among business men. One tractor salesman went out in the country and repossessed six tractors because of delinquent payments. Representatives of the manufacturer were with the dealer, going over the contracts and preparing to repossess the tractors soon to become delinquent and which farmers might use to put in their crops and then let go back. All of this affected the other business men in this town. Farmers were gathering up money to clear their tractors and other business men, the grocer, hardware man etc., were left waiting.

This afternoon I drove over into another county and learned of a 14 cent oats market. The community sale was the largest ever. Seven oats seeders, four hand shellers, nine incubators, plows, harrows, cultivators and every item of farm machinery in varying numbers made up the sale. Some of it was sold because of tractors but much of it for the needed dollars it would bring. I came back to the place this evening much concerned about the future of agriculture and all of the dependent industries. Things and times are certainly changing very fast.

Wed. Mar. 25. 36.

We did odd jobs around the place this morning and about noon drove the truck to the packing plant town, where we marketed hogs. Receiving a price of $8.50 per cwt. The recent wave of selling farm machinery affected us too, but we sold machinery that we had bot last winter to replace our much used machines. We will be able to get along with the old ones this season, the same as we did last year, and all that we sold brot us more money than we had paid; we had bot them low.

Returning home this afternoon I saw the first field work of the season, a tractor disk outfit. The soil appeared too wet for working and I do not think the operator continued very long.

Sun. Mar. 28. 36.

This afternoon we drove to town. The weather was fine today and almost everyone was in town. I saw one of the new "all crop" harvester threshers today and think very well of it from what I could see of the machine on the dealer's floor. When I first heard of these machines I thot I would get one of them for such of our harvesting as we could use it for and to have it in the community. But the present prices of small grain make it out of the question.

Tue. Mar. 31. 36.

Very disagreeable weather today. Cloudy and cold all day. Drove to the county seat this forenoon. Called at the office of the county agent. Could not learn much about when the last of the '35 corn-hog checks would be in. Did learn that community meetings for the new program would begin today. There doesn't seem to be much information available as yet on the new program. The very low price of oats at the present time may cause more farmers to become interested in the new program than otherwise would occur.

Came home thru a light snowstorm that had the ear-marks of a blizzard and late tonight we were having a heavy fall of snow. This will cause many changes in plans for conducting the spring work and the planting of the crops. Livestock, whatever there is remaining in the neighborhood, seem to be getting along in fine shape. Our stock are becoming rather choosy in their rations.

Wed. Apr. 1. 36.

The snow storm that began last evening continued all night and this morning we had from four to six inches of snow on the ground, and it continued to fall off and on all day. Fortunately there wasn't enough wind to cause drifting.

Today I learned of several farmers that have been sowing oats and do not have them covered. This always occurs each year. A few farmers will sow quite early, but this time I am afraid reseeding will be necessary.

Thur. Apr. 2. 36.

Much talk around the sale barn about the soil conservation program. No one seemed to know much about it, but a feeling seems to be developing rapidly that the greater majority of the farmers will go in and become familiar with it as they go along. Every farmer that I talked with who said we would not go in finally agreed to attend his community meeting, register, and see what developed, thinking that this seemed to be the best thing for farmers to do as they are able to size up the situation from this point on the program.[5]

Fri. Apr. 3. 36.

This morning I drove to see neighboring farmers about the men we would elect for committee men on our soil conservation program. In my opinion a new committee should be elected. And some thot should be given to the matter of the location of these farmers in the township.

Our soil conservation meeting was held tonight. At the Center School in the center of the township. Nearly one hundred of our one hundred and twenty-five farmers were in attendance. The county agent was in charge of the meeting. This meeting was not nearly so well conducted as the one attended last Tuesday evening. There the parties, local farmers, who were in charge of the meeting stated that they did not know much about the program, but they talked in an easy, confidential way, giving confidence to the farmers that the details would be carefully worked out and it was a satisfactory meeting. Here, our county agent spent too much time sort of apologizing for the present incompleteness of the program. Nearly all of our farmers who were present, registered and took work sheets. The election was a lengthy affair for just electing three men. They are well scattered over the township. Altogether we are well started on the new program.

Mon. Apr. 6. 36.

Today is April the 6th. Our wedding anniversary and also the date of the day we began farming here at Old Quietdale Farm. I will not mention the number of years it has been. They have been very busy years. And I still think no other place in the world could have brot the satisfaction to us that this piece of land has. We have had our ups and downs and will continue to have them. The world, nature and the weather etc. have not always been kind to us; sometimes cruel and almost very cruel, but the land itself has been good to us. Very kind in fact and we love and respect every foot of the farm. If the weather, markets, folks we associate with etc. would be as fair with us as the soil has we would have life very satisfactory.

Was in town this afternoon. Quite a few farm folks in town too. Our County Soil Conservation Committee will meet Wednesday for a school and I think land appraisals will start Thursday. Corn-hog checks are coming into the state now and our county should be in the money soon. Many snow drifts still remain and as I am writing this evening a wet heavy snow is falling.

Wed. Apr. 8. 36.

The season continues very backward. It is common talk among farmers that they expect a very poor crop year this season. The season is late now and there are many indications that it will continue to drag along getting farther behind rather than catching up any of the lost growing time. Then too the seed corn situation is bad. Not much more than two thirds of a crop is expected at

LILLIAN L. POWERS ("L.L.") AND DANIEL L. POWERS ("D.L."). SPRING 1936.

best. Another factor in the situation is that corn planting will be delayed until late because of the generally poor seed. Then the crop may be overtaken by an early frost next fall, resulting in a soft corn crop. Farmers are not unnecessarily discouraged. They discuss these things as they do good crop prospects when they are indicated. A continuation of reduced buying power is expected. One way to get a farmer ''hopping mad'' is to begin talking about the good prices we are having now. He will soon tell you of the things that he does not have to sell at a high figure.

Sat. Apr. 11. 36.

Another farm tragedy occurred this morning. A farmer living about the edge or to one side of our community, a middle-aged man, committed suicide by hanging himself in his barn. He was well known, much liked, a good farmer and had a nice family. He was about to lose his farm to the mortgage company and I suppose that is what drove him to commit such a rash act. I do not know if this is just a singular incident, or if it is still one of the suicides that should be connected with the number of them several years ago, or if it is the beginning of another wave of troublesome loads that finally become more than farmers can bear. In any event it is very regrettable.

Tue. Apr. 14. 36.

Tonight I was able to contact our soil conservation committee man and completed our work sheet. Now I should be able to finish our field arrangements. From what I can learn the program is going over in good shape. Frequently I hear farmers say "why this is something we should have done long ago."

Wed.-Sat. Apr. 15-18. 36.

A busy, worrysome four days. So busy that I cannot write for each day. Trying to get myself out of the jams, circumstances over which I have no control have got me into, and trying to plan to keep out of other and perhaps worse situations have kept me very busy.

More and more farmers are coming into soil conservation. Business men are guardedly inquiring about it too. The weather has been good and much progress has been made with field work. From very wet fields we have come to very dry ones. Cold and dry. All vegetation at a standstill. Some dust blowing and much talk about how to prevent it.

Another tragedy. Two brothers working together about their farm work and one accidentally backing a tractor over and killing the other, then the next day the one who had driven the tractor shot and killed himself. The greatly increased use of tractors will result in more injuries I am afràid.

Sun. Apr. 19. 36.

Another Sabbath day has come and gone. The weather was unusually nice today. I was here at the place all day. With me, the day was a day of rest. I drew a comfortable chair up to a window and spent some time looking out at the farm. But what I saw and thot about was more than just this farm. I saw all of the other farms as well, all of agriculture. The conclusion that I arrived at was that our nation is a careless, thotless nation in so far as our farm folks are concerned. It is true that a great many people are serious in their efforts to better agriculture, but the nation as a whole is not. They take their food and clothing and the source of it as a matter of course.

Tue. Apr. 21. 36.

When I had finished the morning work I drove to learn a little more about the grass seeding requirements in order that I might quite fully do my part of the soil conservation program.

This afternoon we drove to the county seat and I bot a quantity of sweet clover seed. Many farmers were in and about the seed stores. A few were looking for seed corn, but mainly grass seed was in demand. Sweet clover and alfalfa being most needed. The bulk of the seeding is over with now.

Here at the place this afternoon I took our low down seeder from storage and repaired it to use sowing grass seed. It seems to do a better job than any of the other seeders.

Wed. Apr. 22. 36.

Bill had bot a tractor plow some time ago and I went to town and got it for him this morning. One of Bill's horses was sick when I went to town. The vet was there and when I returned from town the horse was dead. This leaves Bill with just one horse. And he does not have any money to buy another one either. He will need to get one some way in time for corn planting. All of Bill's other horses have died during the past year. In town I learned of other horses that were not able to be worked and of some that had died suddenly. The veterinarians are becoming worried about the situation.

Fri. Apr. 24. 36.

As I had planned yesterday we began the spring plowing today. I was not so sure how the soil would work but it plowed better than we expected. We will use both the tractor and the horse plows and make an effort to hurry the work a little faster than we usually do. It is some bother and extra work for us to keep the new plowing harrowed as we plow it but we will try to do this if possible.

A good rain is much needed by all growing things.

Sat. Apr. 25. 36.

Several neighbors have finished their spring plowing and others have not started as yet. More farmers than usual are disking their fall plowing earlier this season, but I am afraid this will cause more soil to blow. Livestock has been reduced in numbers on many farms until there isn't much manure to haul to the fields. On others where there is stock the manure hauling is being put off and the plowing rushed. An occasional farmer has hauled a good coverage ahead of the plow.

Sat. May. 2. 36.

This afternoon we did odd jobs around the farm and tonight we went to town. The lady of the farm had purchased a new rug yesterday and the salesman had wrapped the wrong one and we returned it to town and exchanged it for the one she had bot. The purchase of a new rug does not occur very often and an event of this magnitude must be handled correctly.

All the time we were in town a rain was falling, always a drizzle and sometimes a hard rain, but before we were half way home we found the roadway was dry and no rain had fallen at our place.

Tue. May 5. 36.

Late this afternoon and evening we did something that our family has never done before. We quit work early in the evening and when we had finished the chores we drove to a State Park for a picnic supper and to watch the moon rise and a drive back to the place this evening. For us to do anything like leaving the farm work and worries for a short time, for no reason at all but to enjoy ourselves, is very beyond the ordinary. We all enjoyed it immensely. We have been to this park numerous times before this, but never at chore time and in the evening. It is something to remember for a long time and we plan to do it again if possible.

Fri. May. 8. 36.

Here and there, a few farmers are still plowing. More of them are preparing corn land for the planter. And an occasional one is planting right along. Very few fields are being thoroughly prepared for the planter. Farmers using horses cannot seem to get the work done fast enough to do it properly and those who are using tractors sometimes are desirous of keeping down the expense. Sometimes there does not seem to be any connection between the cost of producing a crop and the market value of it. There isn't much interest in politics or the markets just now. Everyone on the farms is too busy.

Wed. May. 13. ·36.

For some years I have always looked each springtime for the arrival of the barn swallows and they came today. I know then it is a safe time to plant corn.

Salvation Army folks were driving in our community today gathering funds and clothing. This is the first time they ever asked our farm community for assistance. Their cash donations were on an average of 25 cents. I planted corn as long as I could see this evening and will make every effort to crowd the work as fast as possible.

Sat. May. 16. 36.

Today was a bad day in the fields. The weather was hot. The wind blowing almost a gale and clouds of dust from adjoining farms filled the sky. So far this season we have kept our soil from blowing. In fact it never does give us much trouble. But then we have made that one of our major farm operations, to control this.

The season is moving rapidly along. The blue grass is headed, maple and elm tree seeds are falling and I hunted up last years straw hat to wear in the field.

Neighbors are just as busy as we are. Crowding the corn planting and the field work. Young stock and poultry are doing fairly well, tho some folks have lost all of their baby chicks and are buying new again.

Fri. May. 22. 36.

I do not recall that I have ever planted a crop under such uncertain conditions. I mean as to the returns I will receive for it. The crop may be a large or a small one. The price may be a way high, or it may be very low. Any extreme may occur. Perhaps it will be an average year, but it seems to me to hold greater opportunities for unusual extremes.

I went to the community sale in Boone this afternoon. A very small attendance and offering. A rather uneasy and hostile crowd. Some one took our egg case from our truck parked less than a hundred feet from the entrance to the sale pavilion and between two and three o'clock this afternoon. This is the first time I have missed anything in all of my sale attendance.

Our community was visited by a heavy rain this evening. This will insure all of our corn germinating together. Some of it was planted in very dry soil.

Sat. May. 23. 36.

Today was rather a hectic day wherever farmers were together. Especially those farmers who have corn loans. The report is out that the loans must be paid by July 1st with interest and all expenses and on the basis of the 45 cents loaned. Farmers say they were told to prepare their cribs for a two year period and many of them went to extra expense on that account. Then they say that it is too late for them to abandon the soil conservation plan and plant cash crops on the land. Moving all of the 90% of the remaining sealed corn on the market in this short time will force the price under 40 cents. Altogether they are quite worked up about it. However, I do not think many protests will reach anyone higher up because farmers are too busy at this particular time. Also many farmers feel that the speculators will yet realize more on their crop in their speculative ventures than the farmers do as original producers. Other farmers say this is a false report or that changes will be made so the loans will be continued. At any rate it has had a very bad effect on the administration as far as the farmers are concerned.

Tue. May. 26. 36.

I cultivated corn again this forenoon.

The work progressed more satisfactorily today. I was more familiar with the machine and the horses were more accustomed to the work too. Corn cultivation will be rather a mixed up affair in our neighborhood. There will be fields with two or three cultivations and other fields with one and some of the corn will be very late in laying by.

The last of the 1935 corn-hog checks came in today and we went to the county seat to get ours. It was a little larger than we had figured on. Ours was $91.95. This is probably about what the average quarter section farm received. The checks were passed out at the county Farm Bureau office and I noticed that not very many farmers renewed their memberships in the organization at this time. Many of them went directly to the banks.

Vegetation is making wonderful progress. Hay and pasture grass is coming fine. White Dutch clover is in bloom. Oats and wheat coming along well. An occasional field of corn looks perfect. I have just about made up my mind that we will replant some of our corn.

Wed. May. 27. 36.

Cultivating corn again this forenoon and to town this afternoon. The corn loan matter is developing a rather uncomfortable situation among many farmers who have loans. It does not seem to be the money they might make or lose, but rather what seems to them to be a sudden idea on someones part to close the loans.

Their contracts seem to contain something they did not know about. The average farmer did not read all of the loan contract. They remember they fixed their cribs for two years. They have planned for soil conservation and now to put the 90% of sealed corn on the market in 30 days does not meet the approval of some of them. Whether it is a political move or not many are saying they will change their vote.

Fri. May. 29. 36.

There is much dissatisfaction about the closing of the corn loans. One released crib, because of the grade netted only thirty-three cents per bushel. To have continued holding it for a higher market would have perhaps finally resulted in more than clearing the loan. Somehow the idea seems to persist that it was sealed for two years if necessary. And at a guaranteed price of 45 cents.

Sat. May. 30. 36.

A short day of corn cultivation today. Because of the extreme and unreasonable heat men and horses suffered much. Men on the tractor cultivators were enveloped in a cloud of fine dust. I worked the horses and the dust remained around my cultivator to some extent. It clogged the nostrils and made breathing difficult and it filled one's mouth and throat too. I did not walk in any of the small grain fields today, but I am certain that the severe heat injured them to some extent. And there are indications that the heat will continue.

Mon. June. 1. 36.

Quite a busy, interesting, day today. Primary Election was held today. We from our farm did not go to vote until after dinner. At that time parties in charge of the election reported a very light vote.

I walked thru some of our oats fields today and found the ground very dry and hard. Wide cracks in many places. I have walked thru many corn fields, checking the stand, and find it to be quite light. Probably the lightest stand for many years. It is true there are a few good fields. And it is also true that as the corn grows taller the stand appears better than it really is. I have heard that there are communities where the farmers have left off field work entirely and are all busy shelling and marketing corn.

Thur. June. 4. 36.
Now rumor says that the soil conservation payments, mentioned last spring as being placed at $12.00 to $15.00 per acre will be only $8.00 per acre. This and dissatisfaction of present corn loan arrangements is causing much discussion wherever farmers meet.

The weather is a little warmer and the color is coming back into the corn again. Oats and all small grain and hay and grasses continue to do well.

I have always lectured and talked safety everywhere and especially in my own element and today I was looking at a bunch of cattle in a pasture near a railroad crossing and drove my truck on the tracks in front of a train. I did not even look for a train, was looking away from the direction of the approaching train and at the cattle and never thot a thing about a train until I heard the whistle. It was too late to stop or jump out of the truck. My only chance was in speeding up the truck and I made it. The train engineer was trying to stop and other cars and trucks were turning in my direction. But I made it in time. I fully expected to have the box taken off the truck. I have concluded that accidents on railroad crossings must be blamed on the drivers of the motor vehicles who like myself were temporarily interested in something else.

Sun. June. 7. 36.
Nature must be very kind to the corn crop this season if we are to have plenty of it. And a farmer has a sacred obligation to produce, if not abundantly, at least sufficiently for the needs of his nation. Regardless of whether or not the nation appreciates his efforts and regardless of whether it pays him well or gives him a mere pittance for his production.

Mon. June. 8. 36.
As always, when we begin haymaking there are several things to gather up and some changes to be made. Rope and pulleys have been used for some other purpose and must be returned to the proper places. Our hay is of better quality than usual, tho not any heavier on the fields. It is only several years ago that we did not bother to run the mowing machine over some of the fields. There wasn't anything to mow. That was the drought year. Of all of the many things that annoy country people and their livestock, serious drought is the greatest calamity that can befall them. May we never have another one.

Wed. June. 10. 36.
A free movie by the business men in the village this evening brot the largest crowd to town that has been there for a long time. My soil con-

servation base figures came today. On a farm of 160 acres I am given a base of 123 acres and allowed 104 acres in crop. Our oats are beginning to show heads of grain.

Fri. June. 12. 36

More information is out to the effect that the sealed corn loans will not be required to be closed July 1st as was once reported. Latest advice is that the cribs will be inspected and those that will meet three and four corn requirements will be allowed to continue. This will be welcome news to many farmers and some sort of an arrangement should be made to hold some of the corn in reserve until we know just what the crop will produce for this year.

Tue. June. 16. 36.

The weather was warm and windy. A storm gathered in the northwest but it passed and we did not receive any rain. I would have liked to have had a good rain. The wind and heat today was very hard on the growing small grain and the grasses and soon will show its effect on the corn of it continues.

Hail insurance men came in the field today, but I did not take out a policy. Many farmers seem to regard the soil conservation money they expect to receive as a sort of crop insurance of several sorts.

An outfit of men and machinery from Missouri are harvesting bluegrass seed in our community. They are known as "the blue grassers." I have heard that they are paying one dollar per acre for the acres they cover with their harvesting machinery. They have employed a few local men with teams, paying them $5.00 per day. This is quite a temptation to farmers to neglect their farm work.

Wed. June. 17. 36.

The warm, windy, dry weather continues. There is a little uneasiness among the farmers as to the effect this weather will have on the crops at this particular time.

We continued cultivating and finished the back 40 before noon. Then I cultivated alone the remainder of the day. We do not plan to use the tractor cultivator, only when the horses get behind.

A Capper's Farmer subscription agent stopped by today.[6] I rather like to read certain departments of this paper, but as ours runs some time yet I did not renew today.

Sat. June. 20. 36.

Continued dry weather in the wheat producing sections caused a continued advance in all grain prices. Our elevators are offering 62 cents for white

corn. Yellow corn is seven cents under the white. The attention centering
on the dry areas will cause a study of the grain situation in general and
when the crop situation is as accurately known as is possible an alarming
situation will have been found to have developed.

I was in town again tonight and continued to study the business situation,
especially as [it] affects farming and the farm folks. Bacon in our stores
sells as high as 65 cents per pound. It does not seem that farm hog prices
are up in proportion to this selling price.

In politics the new third party, in our vicinity, will apparently have
much support.[7]

Tue. June. 23. 36.

Continued warm and dry. Crops in our vicinity in good condition. We
raked hay and cultivated corn at our place today.

Thur. June. 25. 36.

Grasshoppers were reported to be very thick several miles from this
town. This is much closer than we have ever had anything like this before.

Clouds gathered this evening and a few scattered drops of rain fell. Not
enough to clear the dust from the vegetation.

Fri. June. 26. 36.

Another day of drought and heat until evening when dark clouds gathered, but only a light rain fell. Not enough to settle the dust.

Our field work today was cultivating corn and I thot several times when
I was in the field that there were more chances of injuring the crop than
benefiting it by cultivating at this time if the weather should continue dry
and warm.

Sat. June. 27. 36.

The heat in our corn fields today was very intense. The corn rolled badly.
The pastures are rapidly turning a deep burned brown in color. I do not know
just what effect the heat will have on the oats crop at this time.

I bot a few more mower repairs tonight as it looks very much like we will
not be able to finish plowing the soil conservation fields because of the drought,
but must mow it instead.

Sun. June. 28. 36.

Another warm, dry day. Wherever folks met they discussed the probable
weather conditions. I do not think crops in our community are injured to any
great extent as yet, but we are very close up to the danger place.

Scattered grasshopper and caterpillar invasions are making a showing
here and there. Perhaps our greatest danger lies in a deficiency of moisture.

Mon. June. 29. 36.

We cultivated more corn again today. I still think that if dry weather is to prevail the corn would be in better shape if we would leave it alone.

At ten thirty o'clock this forenoon there were so many hoppers in the corn field that I looked at my watch to note the time of our first general invasion. I do not look for very many of them, or for much damage by them, but they are here, much thicker than I have ever seen them before. Perhaps they are passing, or were blown in the community by the wind. However, if they are the advance guard of hordes to follow it will be bad.

Interest and anxiety continues about future weather conditions. Sealed corn movement is about stopped. This is because of corn needs rather than advancing prices.

Much of the new seeded soil conservation acres will very likely be lost if the dry weather continues.

Tue. June. 30. 36.

We vaccinated the turkeys this morning. It was four o'clock when we started and we had finished by seven o'clock. This made the regular morning work quite late. There are several things about the turkey industry that give the grower much concern at times.

We cultivated corn this forenoon and this afternoon drove to the county seat to see the banker abour our corn loan. All corn loans are supposed to become due tomorrow. Our banker inquired if the corn was in a good crib and when he learned that it was, suggested that we leave it just where it was for a time. I talked with the sheller operator in our neighborhood today and he said that he had shelled nearly all of the sealed corn now. In fact had shelled about all of the corn in the community.

The weather was hot and dry again today and all grain markets continue a little higher. Our small grain is beginning to suffer. There isn't any moisture in the soil to sustain them and the intense heat is turning the heads white rather than ripening them in the natural way. Our horses continue to work despite the heat. We water and rest them frequently and work them short days.

Wed. July. 1. 36.

I went back to town again this morning. I went to the courthouse to learn about the tax situation for I had been hearing many complaints about it and I learned that there has been a very substantial increase in all taxes. I was unable to learn any definite cause for the increase or what useful purpose was being made of the extra money. Of course the county treasurer's office is only a collecting agency.

The hot dry weather continues. Our pasture is holding up well. However, the corn and oats are beginning to show the effects of the heat.

Many rumors are going the rounds as to the present business situation and the drought and one hardly knows what to believe; however, the old saying that wherever there is smoke there must be some fire very likely holds good.

There were more grasshoppers in the fields today [than] any time yet so far this season. Here around the place caterpillars have devoured more vegetation than the hoppers in the field.

Thur. July. 2. 36.

Corn cultivation was our field work again today. We work in the fields and look around the landscape. What is going to become of it. The corn plants look thrifty but we know they are not growing the foundation for a crop that they should. Sufficient moisture is not available, tho there is abundant heat.

I attended the community sale again this afternoon. The usual offering of livestock came thru the ring, but it went out at substantially lowered values. The heat in and around the sale barn was intense. It had a depressing effect on all. A common expression among the farmers was "Well this will finish the oats." Many farmers in that community were cutting and storing their oats crop for whatever hay it might make, believing that the heat and drought had ruined it for grain purposes. The farmers in attendance at this sale were a serious minded crowd of men. Market news was closely watched and discussed. All grains advanced again today. Small groups of men were in serious discussion as to the best way to meet the coming situation. They are bewildered and uncertain now and will be for a day or two. But plans and leadership will come. Many are hopeful of rain in the next 36 hours and all of them did not think anything like this could happen again. The corn crop is still safe for a few days. Men told me they had been feeding their pastured stock for several weeks, but were keeping it quiet so the bankers would not know it.

Fri. July. 3. 36.

Last night was a hot uncomfortable night. Everything around the farm was restless and fitful. It was a beautiful night outside, with a clear sky and a pleasant moon. But the stock in the pastures could not be contented, and worried and wandered back and forth. The places that had been comfortable before were not now.

We cultivated more corn and brot the tractor cultivator from the field this noon and will put it away until next year. We may do a little more cultivating with the horses.

This noon I drove to another sale barn. There I found a smaller offering of livestock and buyers than usual. As yesterday's weather and market news crowded out all other interests. Local and National relief plans were mentioned and discussed. It was agreed by many that our oats crop had had some chance until this afternoon. There isn't any available moisture in the soil for a short rooted crop like oats and the intense heat is wilting, not ripening it. Dealers report stocks of binder twine as being reported to be moving this way from the west where it will not be needed. Also grasshopper poison is being returned. Heat and drought have destroyed crops ahead of the hoppers. This afternoon I drove north into other counties and found conditions the same, and in the towns and cities business men were much interested in how far the situation would develop into disaster.

Sat. July. 4. 36.

The great National Holiday today. We did not celebrate. But spent the day around the place. The weather was very warm. In fact hot, and hot winds blew this afternoon. In a day or two we will see burned blades on the corn. This afternoon was a very bad one for the oats also.

This evening we drove to a park and spent the evening visiting with folks who had gathered there. I met a number of farmer friends. They are trying to get their oats fields made for hay as fast as possible, thinking that the returns will be greater that way than to allow them to continue to dry up with the possibility of no production at all. The livestock situation is a topic for much discussion, especially cattle. I met a Federal vet tonight who says the Government will begin buying drought cattle next week.

I heard today of 500 farmers gathering in their county seat town where they expected to receive grasshopper poison. They had been looking for it for a week and it did not come in today. No rain in sight tonight. Both barometers indicate continued dry weather.

Sun. July. 5. 36.

Not much to write about today. The drought and heat continues. Dry hot winds were blowing from the southwest again this afternoon. This was very bad for the corn and especially bad for the oats. It seems to me that the potato crop here must be entirely dried up.

Late this afternoon I drove the truck to my brother-in-law's place and borrowed enough iron pipe to reach from the windmill pump to our garden and will pump water on the garden all night. We do not expect to be able to water all of the garden, but may save some of it. The vegetables that will be most useful for canning purposes.

Not a chance for rain tonight. We are studying the problem of what is best for us to do with our oats crop.

Mon. July. 6. 36.

The hot dry weather continues. I looked over our oats fields and decided to harvest them. D.L. cultivated corn all forenoon and I worked over the grain binder.

This afternoon we began cutting. I found the grain just a little different than any I had ever handled before. Many of the straws had been cut off by grasshoppers. Also many were broken over. I do not know what threshing returns will be. The ground in the oats field is very dry and hard. The binder jolts along and is rough riding. Great cracks are in the ground. I could drop my pliers down out of sight in many of these cracks. I am pulling the binder with horses. Almost every binder in the community is in the fields today. Nearly all of them drawn by tractors. Whether oats appear green or ripe, they are being harvested anyway. Many farmers have a feeling they will be as well in the shock as standing. Also the general opinion seems to be that one would do better not to cultivate any corn now. Our windmill pumped into the garden all day today. We are making an effort to save the garden if possible. Reports are that all grains advanced again today. Many farm fires are occurring. Both buildings and fields are burning. No rain in sight tonight.

Tue. July. 7. 36.
Another day has gone by adding one more to the heat and drought. With the exception of several small clouds in the far northwest the day was clear. We had hopes from these clouds, but they were small and instead of more and larger clouds gathering, the small ones disappeared. The wind was blowing all day and our windmill pumped steadily all day.

We harvested oats again today. They do not seem to be ripening in the usual manner, but we will continue to harvest them anyway. There isn't a particle of moisture in the soil to further their development. We have concluded that they will thresh some quantity of grain. How much and of what value is problematical. Many grasshoppers are in the fields. It seems they are increasing in numbers each day. I do not mind them only when they fly in my face.

The twine I am using is left from last year but I must buy more tomorrow. I oiled my windmill today and it seemed that the surrounding fields did not have their healthy growing appearance that they usually have. It is beginning to be conceded that the drought is affecting the corn crop. There is much talk that the stalk growth while as tall as usual is not a substantial growth.

Wed. July. 8. 36.
The weather continues very hot and everything is becoming more dry each day. Dan says a farmer in his neighborhood attempted to burn a small patch of quack grass and the fire soon was out of control and it required everyone in the community to get it out before it would burn many hay and grain fields. As it was a 20 acre field of hay was burned.

I harvested this afternoon. Hot and dry and dusty. The horses worried considerably and I watered them frequently. Near evening, looking toward the sun I noticed "sun dogs." Usually they indicate a severe wind and electrical storm at this time of the year. I am still hopeful of rain in time to benefit some of the corn. There are fields that very likely are beyond help now.

Thur. July. 9. 36.

The oats continue to dry up and I think I will finish the harvesting as soon as I possibly can. The earth is dry and hard and many large cracks are appearing in the stubble field. Any tools that I carry on the binder may be dropped down in these cracks. I tied a string on the handle of a 12 in. crescent wrench and lowered it down a crack. I will not mention the distance. Some things are better left unsaid, even about a drought.

A farm house burned today. From another nearby community comes the story of a 20 acre field of oats being devoured by grasshoppers during one days time.

At the sale I found a cautious, worried crowd of buyers. Hogs that sold at $24.00 to $27.00 two weeks ago went thru the ring today for around $16.00 to $17.00 and what were $7.00 to $8.00 pigs two weeks ago were $3.00 to $4.00 pigs today. Other livestock in proportion.

Fri. July. 10. 36.

I finished harvesting this forenoon, about the middle of the forenoon. Then I went to town and learned that the grain elevator was paying 78 cents for corn and 36 cents for oats. I have my corn yet, my sealed corn. If it should advance enough it would pay the corn note and my other notes at the bank, so I am holding and trying to watch the market. Other than twine, my harvest cost me ten cents for small staples for canvas slats.

A cigarette thrown from a passing car is blamed for starting a fire that burned over quite an area of pasture field today.

This afternoon I drove to the capitol city. I spent the afternoon visiting with an aged Uncle and we drove home late this evening. In spite of the heat and general discomfort that goes with it I enjoyed the trip as a change from the daily grind. Everywhere we went we found drought and heat and injured and ruined crops. I saw a hundred acre field of corn in our neighborhood today that is ruined by heat. It is beginning to tassel now and standing two to three feet tall. And it is all very weak and spindly and turning brown on the bottom of the plants.

Sun. July. 12. 36.

Today is the tenth consecutive day that the thermometer has registered above the 100 mark. The reading today was 105. There are newspaper and radio forecasts of approaching rain but I fail to see any indications of them. Our various weather instruments do not indicate any rain in the near future.

I walked around the place some today and wherever I went I found the grass crackled and crushed under my feet as I walked. I can recall several times when we had a severe drought but I do not recall anything as bad as this.

Our livestock and poultry are standing the extreme heat much better than the growing crops. This present drought covers more territory than perhaps any other on record.

Mon. July. 13. 36.

And still the heat continues to increase. The thermometer, early today, surmounted yesterday's record, reaching 106. A few light clouds came in the sky but they did not appear to be rain clouds. About the middle of the forenoon a slight breeze came from the northwest, but it soon passed and the heat seemed to be more intense after that. Apparently there isn't any rain in sight for several days more at least.

I trucked hogs to the packing plant for Bill. The plant was paying $8.90 for butcher stuff today. Around town there were many indications that the town people were not standing the heat as well as we farm folks do.

Nearly all of the grain harvest is completed and I saw one threshing rig in operation today. The corn market eased back to 75 cents today. Our mail brot letters and bulletins on grasshopper control. So far our county has escaped serious ravages by these pests.

Tue. July. 14. 36.

More heat again today. The 12th consecutive day above 100 degrees, reaching 108 this afternoon.

I drove into the city this afternoon and it was the most disagreeable trip I ever made as far as the heat was concerned. At every filling station I found travelers with burned faces and some of them nearly overcome with the heat.

Fires continue to be one of the great hazards of the dry weather. Today I saw a place where a fire had apparently started by the highway and burned across a pasture and several rods in a corn field before it had been extinguished. There isn't very much moisture in some of the corn plants. It seems strange to think of what seems to be a green corn field burning.

Many farmers are attaching gas engines and pump jacks to windmill pumps, as many pumps are required to work continuously. I was in two towns today where warnings were displayed to conserve the water.

Wed. July. 15. 36.

Another of the very warm days again today. This is the 13th consecutive day that the thermometer has registered above 100, reaching 108 at three o'clock this afternoon.

Here at the place we worked to keep the livestock as comfortable as possible. Crops continue to dry up and our prospective yields diminish.

Thur. July. 16. 36.

This is the 14th continuous day with a thermometer reading above 100. And no relief in sight.

I mowed along some of the roadways and on a part of the soil conservation field this forenoon.

This afternoon I drove to the community sale. I found a very light offering and a small attendance of buyers. All values very much lower. Pigs that would have sold for up to eight dollars per head three weeks ago went out of the sale ring at three to three and a half and sows were selling at sixteen to eighteen dollars per head instead of twenty four or five several weeks ago.

Cattle values were not quite so low.

The seriousness of the drought is becoming more real each day. One does not seem to learn much about it from the papers, but travelers, truck drivers, etc., who are moving about the country give various reports of it. And one may make all of the allowances and deductions from these stories that he cares to make, and after that they are still plenty bad enough.

Fri. July. 17. 36.

The drought and heat continues. Each day it continues more fields of corn are injured beyond any possible relief.

Late tonight it seemed that there might be some change in the weather and that a cold wave might come. Lightning flashes in the distance and static on the radio gave hopes for rain.

Sat. July. 18. 36.

We did have slightly cooler weather this morning. And a very light sprinkle of rain. Not enough to wash the dust from the parched plants.

Sun. July. 19. 36.

The greatest corn crop disaster that our country has ever experienced is upon us. It may take some little time yet for the fact to become apparent to all of the citizens to grasp the idea and some of them may miss it entirely, but all of them will know eventually that something is wrong some place.

Clouds gathered in the northwest this afternoon and a very light shower of rain fell here. Hardly enough to wash the dust from the shrinking corn. And the thing we have been worrying about happened. We knew that the corn plants were not developing normally and when the wind that accompanied the sprinkle of rain came along it blew the corn over nearly to the ground. Perhaps the flattest I ever saw corn blown down. Many hills of corn were hardly rooted and these tipped right over, loosening the plants in the dry soil. Many stalks are broken off several inches above the ground too. The wind damage to the crop amounts to much more than the small amount of moisture benefits. Our oats shocks are blown every which way, but this is not a damage, just an inconvenience.

Mon. July. 20. 36.

This morning I went out over the farm to note the crop damage by the wind and also noticed that the rain had not amounted [to] enough to settle the dust. I decided not to rebuild our oats shocks because we will be threshing soon and chances of damage from rain are almost none at all.

Later in the forenoon I drove around the community and to town. Common opinion seems to be that the amount of down corn will reduce the yield by about seven bushels per acre. That is, it would if it was an average crop. Drought reduction estimates are commonly placed from forty to sixty percent. Farm folks are becoming disgusted with the continued radio and newspaper crop reports telling of the slight damage and that rain would cause recovery of the crop. Also they are tired of the crazy speculation going on in the grain markets. It is bad enough when the farm income is up to standard, but when famine and starvation are so close it is a serious matter. Imports of farm products are becoming a source of much discussion too.

Late tonight I drove ten miles to a farm fire. A fine barn and much valuable feed was destroyed. Matches and smoking are becoming a very serious problem.

Fri. July. 24. 36.

We threshed at our place today. The weather was warm and a fair wind was blowing and the heat did not seem oppressive.

We use eight bundle racks and from two to three men care for the grain. The machine has a crew of two men. It is a gas tractor outfit. They charge two and one half cents per bushel for the work. We have 33 acres in oats and we finished by the middle of the afternoon with a total of 1361 bushels. I did not figure it out, but someone said it was a yield of 41 bushels per acre. All of them were put in the bin but a load of 120 bushels were sold in town for 32 cents per bu. All of the straw was blown into the barn. This required setting the machine twice. And I worked in the mow a good part of the time. The straw seemed very dusty. It seemed to be a sweet tasting, greenish dust. I never saw anything like it other threshing years. We gave our threshers dinner here at the place and I think they enjoyed it and that we liked to serve it. It was not an expensive dinner either. And so another threshing has come and gone and another oats crop cared for. I was much surprised at the yield and sometimes wonder if the grain measure on the separator was working properly. I do not see how our fields could have produced that much grain under these conditions. Of course our oats fields were the lowest land on the farm this year and they were rather protected from the warm winds too.

Sun. July. 26. 36.

When we had finished the morning work this morning we drove over into Carroll county to see the Roy V. Copp farm and herd of Chester White hogs.

I have been thinking of purchasing a gilt for a fall litter and wanted to see how the crops were over that way too.

We had thot things were dry here but we found them much more so there. The pasture fields were absolutely bare and black looking. All cattle and horses are fed hay. The corn plants were from a third to a half as tall as they should be and I cannot see how there can possibly be any corn to husk and I doubt if there will be any for fodder. There seemed to be a hundred grasshoppers for every one we have here. Driving along the pavement we were never out of sight of a place where a fire had been burning. Apparently matches and cigarettes had been thrown from passing cars. Many of these fires had been confined to the highway but some of them had burned diagonally across oats fields, hay meadows and pasture fields and some of them it seemed had nearly destroyed whole sets of farm buildings. Tractors and plows attached were ready on alomst every farm for instant use.

Wed. July. 29. 36.

Our grain markets reached new tops today. $1.00 for white corn and 90 cents for yellow. 35 cents for oats.

It is becoming quite generally admitted that the drought is the most serious that our country has ever suffered. In Carroll and western counties farmers are said to be harvesting their corn fields with grain binders because regular corn binders will not handle the short crop. We have corn in our community that will not make fodder. And many of the fields that we had hoped a rain would benefit are being found to be fields of almost all barren stalks.

A crop expert says that the state of Nebraska alone will take all of the grain that our state might have to spare. I talked with a farm hired man from Greene county and he says their corn crop won't make 15 bushels to the acre. Grain elevator men here are worried because they believe farmers here are selling their oats too close.

I repaired some of the barn roof this afternoon. It should have a new roof, but because of the drought it will be out of the question this season.

Thur. July. 30. 36.

The dry, discouraging weather continues and each day seems more hopeless than the one just passed. I hitched to the mower and mowed along the fences around the stubble fields. The few weeds and scattered grass are quite dry and dead but I thot the field would look better this way.

Fri. July. 31. 36.

The weather was slightly cooler this morning and throughout the day was not quite so warm as some days have been. Streams are dry and pastures are bare. We cut off corn for our stock. There is much talk about baling straw and saving it. My understanding is that there has been a fair crop of straw everywhere.

The railroad shops laid off 90 men today and many section men did their last days work today.[8]

Sat. Aug. 1. 36.

No change in the weather. Markets continue to advance. No. 3 corn up to $1.02. Oats 49 cents. One elevator firm has 80 elevators in Iowa and have discontinued selling corn at all of these elevators but one and the manager expecting a message at any time closing this one. These 80 elevators average a storage capacity of 70,000 bushels and they are all nearly all filled. Their contents insure a supply for the Cedar Rapids mills for one year's operations.[9]

I went to town this afternoon and found the office in the court house and many others closed for the afternoon. Quite a few farmers were in town. All agreed that rather difficult times were ahead, but each one had a different plan for meeting the situation.

I went back to town tonight and did not find nearly as many farm folks in town as there were this afternoon. Business men are becoming worried about the future. Business has almost stopped for all except the grocery and food stores. Many farmers freely admit they will not be able to make interest and similar payments.

Sun. Aug. 2. 36.

Just another average Sunday today. The usual company came and the usual things happened. The heat and drought continue. We did talk about driving west to look at some of the relief counties, but did not.

This evening we attended church in the village church and when we returned home from church we saw a small fire along the pavement and started to drive in that direction and a few moments later we saw the second fire. As we passed Bill's place we called him out and drove on to the two fires. They were near the village and burning toward the town with only the railroad track between and a rising wind could easily cross that. Arriving at the fire we found the section foreman already there and Bill went to work with him and I drove to a nearby farm where the hired man was already backing the tractor to hitch on the plows. From there on I drove ahead of or beside the tractor to light the way. A pasture with a heavy growth of dead, dry grass was burning and by the time the tractor had crossed the fields to the point of the fire, a small army of men with shovels, grain sacks, water buckets, hand extinguishers etc. had the fire under control. It was surprising where so many men came from so soon. This was about nine thirty to ten o'clock in the evening, before many of them had turned in for the night. Then too seven highway fires were reported between nine and ten o'clock between Ogden and Grand Jct., and some of the men seemed to be doing volunteer fire patrol duty along the highway.

Mon. Aug. 3. 36.

Grain markets were higher today. Corn at our elevators was up to $1.06 for No. 3 yellow. Perhaps I should have sold our corn, but with the difficult times ahead it is natural that I should wish to realize as much as possible on it.

I loaded one of our water tanks in the truck and took it to town to have it repaired today. It was a metal tank and has been leaking considerably of late. I also had mower repairs made instead of buying new parts. It will be get along with the old thing for some time to come. "Going Scotch" will be popular with lots of farm folks and "going without" will be forced on some of the city folks.

I mowed some in the hog pasture today. There were a few scattered weeds of the deep-rooted sorts, and I mowed over the whole pasture to get these few. Everything else in the pasture was dead and dry.

A very light shower of rain fell this evening. Not enough to settle the dust.

Tue. Aug. 4. 36.

The grain markets broke badly today. Scattered showers were reported as the reason. There is becoming much speculation in the grain business. The producers and the consumers and their rights and interests are entirely ignored in the matter. Cash corn went down seven cents. I am still holding mine in the belief that when the crop disaster is fully known that corn will sell at around $1.25 cents per bushel. That amount will pay my notes at the bank and my Land Bank payment. Then too I am testing my corn to see if it will do for seed. In a day or two I should know about this.

I drove to town this afternoon. Quite a few farm folks were in town this afternoon. Farming is changing again. So many farms have passed into the hands of banks and insurance companies that the people living on them do not have anything to do but the actual farming of them, and because of the drought they won't have much to do but the fall plowing. Corn husking will be a short job this year.

Wed. Aug. 5. 36.

Markets a little lower again today and the drought effects much more apparent.

I did odd jobs around the farm and more repairing the buildings. Some building materials go out to the farm each day. Some repair work and a few jobs of remodeling. Farms owned by insurance companies and banks are receiving more attention than those owned by farmer-owner operators.

There isn't much field work being done these days. The ground is too hard and dry for working in the fields and a few spreaders are in use.

I am having trouble trying to test seed corn. I cannot keep the grains moist enough to get them to germinate. The corn itself is so very dry and the air is dry too and almost before I know it the test box is too dry for good testing.

Thur. Aug. 6. 36.

We had a surprising change in the weather this morning. Sometime last night a fog came down and this morning the air was quite damp. The forenoon was well advanced before the fog had disappeared. The slight amount of moisture was much appreciated by the corn plants. Tho many of them are past much help.

Fri. Aug. 14. 36.

Clouds covered the sky this morning and we expected the long awaited rain, but it turned out to be only a light sprinkle.

We did odd jobs here at the place and I watched the radio market news and went to town several times. I was thinking about selling our corn but the market continued erratic and lower and the custom sheller was out of operation because of a breakdown, so I did not sell. I am hopeful that the market will be up again in a day or two.

Mon. Aug. 17. 36.

Studying and trying to figure out the corn market and what the future will do has been my principal job the last week or two and this morning I found the corn market working slightly higher. I need to realize all I can from the old corn and yet must determine when the right time comes to sell.

The weather continues warm and dry.

Tue. Aug. 18. 36.

As the drought continues and it is much too dry to plow our stubble fields and as there are several patches of some kind of weeds that might give us trouble, I hitched to the mower and clipped them. Several times this forenoon I stopped work long enough to dial the radio for market news and found the markets working higher again today. Later in the forenoon I drove to the village and learned that they were paying $1.07 for #2 corn.

This afternoon I drove to the county seat, to learn if anything had been done by the Federal Land Bank in the way of easier interest payments and found that no changes had been made as yet. Many farmers were in town and they were a perplexed lot of men, to know what and how to do things.

This evening near chore time I drove to the village and a new straw baling outfit had moved into the community and I stopped to see it at work. It was quite efficient. One of the workmen was smoking a cigarette and several times brushed it on the machine to knock off the ashes. And we have had four baler fires here the last ten days.

The weather was very warm today and especially this afternoon. Hot winds were blowing. Many folks said the hottest day they ever experienced. This wind, in addition [to] burning the corn plants, burned the shucks on the ears. It was very bad. Late tonight a rain storm passed south of us.

Wed. Aug. 19. 36.

The corn market reached $1.07 at one time today and I should have sold then. The interval of high was so short because a wire changed the market again. Now it is backing down and I am wondering if and how soon it will reach that figure again.

Thur. Aug. 20. 36.

I did some work in the field today. It has been a long time since I worked in the soil. I replowed some of the soil conservation ground. One can hardly call it plowing, but that was the implement I used. The earth is so very dry and it just kind of pushed around in front of the plows. When I tried to harrow it, rolling the hard lumps and clods around was all that I accomplished.

The corn market is still going down, having worked down five cents today. Perhaps I should have sold mine. However, by waiting until I know if we have any to husk I will be sure of feed.

Fri. Aug. 21. 36.

The corn market did not change much today, remaining about the same as yesterday. There isn't much activity in the grain markets here now.

Governor Landon's special train passed thru our community today and just about every one made an effort to be at one of the stopping places.[10] It was a serious and very friendly crowd that met the Governor where I was. And I believe he made many new friends.

A light shower of rain came to our neighborhood this evening. Just enough to cool the air, but not enough to aid vegetation.

Sat. Aug. 22. 36.

Local elevators were paying $1.08 for corn today. The market was draggy and I did not sell because I thot the end of the week has always been the low time and next week should see some improvement. There is some talk about South American corn coming into the country and all farmers are very indignant about that. Even feeders who know they must buy all of the corn they feed and the ordinary farmers who will not husk any corn resent the idea of our not having made allowance for a possible crop failure and giving our market to another country.

We have city company this evening for the week end. Two families. One from Newton and the other from Waterloo. One party works in a washing machine [factory] and the other in a tractor factory.[11] Both of them remarked that South American farmers would not buy our washing machines and tractors but if the imported corn were kept out farmers here could more easily buy our goods.

Wed. Aug. 26. 36.

The corn market was very erratic today. I went to the village elevator and spent much of the forenoon there. My radio is temporarily out of commission and I used the radio in D.L.'s car for the opening market. Then in the village I watched the market climb to a dollar and four and one half cents per bu. I am not attempting to sell at the very top but just to get enough money together to get certain obligations cared for. So I thot this market would come quite close to doing it. The dealer suggested the better the grade of my corn the more money he could pay for it and I returned home for sample ears of crib run corn for him to determine the grade. When I arrived home for the sample ears I found the cows were in the corn field (the second time for this season which is a better record than the average farm here) and by the time I had returned the cows to the pasture field and taken the corn to the village the market had dropped several cents and we decided to wait a day or two for a possible better market. It is just one of those things that do happen sometimes. All others of the family were in Boone to see the Achievement Show Parade and there wasn't any one to take the sample corn to the elevator or to drive the cows from the corn.

Thur. Aug. 27. 36.

The heat and drought continue to take their toll. In an adjoining community, back in harvest time several farmers were overcome with heat and taken from the binders to hospitals. Some of them did not recover. A farm sale of one of these victims was held today. The livestock and farm machinery as well as the crops yet in the field, brot good prices. The widow could not continue alone and she and the small children will move to town.

I was in a town, on a trunk line highway, this afternoon and counted thirty-one hitchhikers lined along the street and out into the country, all looking for a ride. Mainly they were young men. I picked up one of them whose home had been in a western Iowa town. He had been away to New York state for eleven months. Said every kind of business was at a standstill there. I tried to rather prepare him for some of the things he [is] sure to meet up with when he arrives back at the old home.

Corn is down to 96 cents today.

Thur. Sept. 3. 36.

Today is a big day for the State of Iowa. With the President and Governors in the State to hold a drought conference, quite a few planned to attend the capitol city.[12] We did not go and only several farm folks from our community did go. A great many farm people are planning more carefully, working more, spending less and trying to save more than they have been doing in recent times. In other words going Scotch with a vengeance. And most of the folks in our community would like to see the President and all of that, but it doesn't fit in with their program.

We worked on the soil conservation ground today, harrowing and rolling it preparatory to seeding as soon as possible.

While rains are being reported all around us and the drought isn't mentioned so much anymore, the water problem is a very serious one on many farms. And all kinds and sorts of arrangements are used to supply the farm stock with water. Much of it is hauled. The longer hauls with tank trucks and the short ones with a barrel on a stoneboat. The practice of

moving young livestock and poultry to clean ground has done much to
provide some means of hauling water on many farms.

Several times during the night last night I was awakened by the sound
of rain falling on the roof or driven sometimes by a steady wind against the
side of the house. There wasn't any thunder or lightning and only several
gusts of wind. After that just a slow, steady rain, the kind that does not
storm down the corn. This morning it was evident that at least two inches
of rain had fallen. Now the drought is broken, however, it will be some
little time before the grass will become green again. And unless we have a
very late fall not much vegetation will be benefited by this rain. It will be
possible to begin field work now. It is rather hard to understand the whims
of nature sometimes. These weather extremes are often very cruel to some
communities, just as they are beneficial to others. I do not know how long
it has been since we have had a real rain. So long ago that I have forgotten
all about it. From now on I look for a continuation of sufficient moisture
for all needs.

We plowed some on the soil conservation fields today and went to town
tonight. The grain market is at a standstill just now, corn hovering around
a dollar. I think I shall sell ours before the market opens next Tuesday
morning.

When I had finished the morning work I went to the radio and listened to
the opening grain market quotations. Then I drove to the village elevator
and found that I could sell the corn for Saturdays price which was $1.00 ½
for grade No. 2., and I thot by sorting the corn some as we shelled it we
could make it this grade. Crossing the street to the garage I met the sheller
operator and learned that he could shell the corn this afternoon if I sold it
now. And going back to the elevator I sold 1300 bushels. Last December we
had sealed 1375 bushels but I thot it would shrink perhaps a hundred bushels
anyway.

The sheller came after dinner and we began shelling. Our truck and the
truck with the sheller hauled the corn to town. Several neighbors whom I
had exchanged work with and who now owed me work shoveled the corn from
the crib and during the shoveling sorted out any discolored ears of corn they
happened to see.

We finished the shelling in several hours and the sheller operator and
some of the neighbors thot there was around fifty bushels of corn that we
had thrown out as unfit to sell and corn that might spoil the grading of it.
Later in the afternoon when I drove to the elevator I learned that only 1025
bushels of shelled corn had come in. That quantity, with the fifty bushels we

had thrown out made a total of 1075 bushels. Leaving a shortage of exactly 300 bushels. Certainly it had not dried out and shrunk that much.

Wed. Sept. 9. 36.

This morning I went to the elevator to settle for the corn and I was still trying to account for the shortage of 300 bushels. The elevator man said no other crib had a shrinkage like that. Thinking back to December when the corn was sealed I remembered that the folding ruler that the sealer used had one or two lengths broken off of it. I decided that he had made an error in measuring the crib. I recalled that I had left the measuring etc. all to him because I was busy figuring out a way to raise money enough to meet the old mortgage and clear up the Federal Land Bank Loan. So now it was apparent that the sealer was in error that amount and I had borrowed money on 300 bushels of corn that I never had. It made a low rate of interest anyway.

I have watched numerous sales of corn at country elevators and never liked the grading system they used. My own corn this morning was made to grade so that I was paid 97 and ½ cents for it instead of the 1.00 ½ I had rather expected to get for it.

Later in the forenoon I drove to the county seat and instead of paying off the corn loan, the loan I owed the bank and most of the Land Bank payment, I had to borrow all of the Land Bank and renew part of the local bank loans. The banker was very obliging and helped to plan to eventually get these things all paid.

This afternoon, here at home, I began to plan again for the things I hoped to do to repair the buildings and get them painted and to plan for another year for I can get thru 1937 now. I wish everyone could do this well.

Sat. Sept. 12. 36.

The weather cleared this evening and we drove to town, but did not find many farm folks around town. It was "bank night"[13] at the largest theater. Admission was 36 cents and the place appeared to be crowded. I did not go in and I do not think many farm folks made up the audience.

Tue. Sept. 15. 36.

Today was a rainy day, all day. Light to heavy showers. Sometimes accompanied by thunder and lightning. It is the first rainy day for months. With continued normal rainfall there will be sufficient moisture for all needs now. However, if additional rains do not continue soon this wet spell will all be absorbed by the lower subsoil.

It was interesting to watch the livestock and the poultry. Much of the younger stuff did not seem to know just what the rain was. There have been several small showers but they did not add much to their discomfort. The poultry drooped thru the day and went to roost early.

I did many small jobs around the barns and in the shop. Much of our machinery should have been in storage before this rain but we have been delaying because of the probability of using some of it.

Wed. Sept. 23. 36.

Market prices of cattle and hogs continue to work lower. Scattered local political gossip just now seems to indicate much interest in Governor Landon. The only Roosevelt supporters are the soil conservation committee men and a few on relief and in office, rumors say.

Thur. Sept. 24. 36.

We kept the plow going this forenoon and this afternoon D. L. threshed sweet clover seed for a neighbor and we others of the family attended the Greene County Fair. It is the only County Fair left in our neighborhood and I have regretted to see them go. On the whole the fair is quite good considering how severely the drought has affected this county. The few cattle on exhibit were in good condition, but the horses were thin. Not so many head of live-stock, but showing better breeding and needing more feed seemed to be the livestock situation. The attendance was quite light and tomorrow is planned for the big day. A new car is to be given away tomorrow evening and many folks, both town and country people, are acquiring free tickets with purchases of merchandise and expect to be in attendance tomorrow and tomorrow evening to see who will win the car. The admission charged seemed to be rather high. Forty cents at the gate and twenty-five cents for the car. Twenty-five cents to the grandstand and fifteen cents for the bleacher seats. The same rate again this evening. Part of a carnival company made up the bal-ance of the attractions for a small midway.

Thur. Oct. 1. 36.

With the beginning of this month we are entering the last quarter of this year. If the weather should be favorable much of the corn crop will be cribbed this month.

Since our pastures are becoming green again I have been thinking that we should have more young cattle in them. So I drove in to see the banker and he favored the idea. He suggested that I go out and buy whatever I wanted and come in and we would fix up a note to cover the purchases.

This afternoon I attended a community sale and looked over the calves and young cattle very carefully. Then when they came up for sale I bot five head of them. Several white faces, several reds and a roan. I paid $47.50 for them. And loaded them in the truck and was back here at the place at dark.

I talked with many farmers at the sale barn. One is husking, with a yield of seven bushels per acre. Another has thirty-six bushels from 20 acres. Tonight I attended a small-town carnival, just out of our community. It was a very

trashy affair. Mostly beer joints and girl shows. Most of which admitted they
were kicked out of former locations.

Fri. Oct. 2. 36.

I drove over to a neighbors this morning and bot a white face calf from him,
paying ten dollars for it.

Then I hitched to the gang plow and plowed for several hours. The plow we
have had resharpened in the shop is working properly now. Also our soil is
becoming dry enough that the plows can handle it again.

This afternoon I went to another community sale and bot five more calves.
I paid $49.50 for these. I suppose that I might find them by driving over the
country, but this way I have a choice of several and the opinion of the other
buyers as to what the stuff is worth. I bot well bred calves but thin ones. The
packer buyers took the choice ones at prices farmers could not hope to pay.

Sat. Oct. 3. 36.

Finishing the morning work I drove to town to see the banker and as I had
checks past $100.00 for the 11 head of calves and young cattle I had bot I gave
him a note for $125.00 and will buy several head more. I plan to weigh them in a
day or two and then keep them for a time at least.

This afternoon I did odd jobs around the place. Picked up the corn in the
fodder field and plowed oats stubble. I have all of the lands started now. The
soil worked perfectly. Last summer when the earth was so very dry and great
cracks appeared, it seems that this cracking loosened up the soil, very much as
blasting would have done. Then the rains mellowed it and I am much interested
as to the kind of a crop that will be produced next year.

Sat. Oct. 10. 36.

We did quite a variety of work today. D. L. husked corn and I worked with
him a part of the forenoon. This was the first time I have been in the field this
season and I have been trying to prepare myself for what I expected to find.
However I had hoped that it would not be too bad. We husked fifteen rows
eighty rods long and hauled the corn to town where it weighed 20 bushels. We
thot it would be close to 30 bushels by wagon box measure. The grain elevator
bot the corn, figuring 80 lbs. to the bushel and paying 84 cents for it. I usually
like to sell a first load to check up on the corn.

Fri. Oct. 16. 36.

This morning I drove over into another county to a tractor parts company,
where they wrecked a going tractor to give me a used gear for the price of $2.00.
We hear a great deal about how business has improved and how everyone's
income has increased, but this salvage dealer seemed glad of the opportunity
to break up a serviceable tractor for so little money. Of course he will very
likely sell other parts.

Tue. Oct. 20. 36.

I met a number of political friends of both of the leading parties. It seems that about everyone has decided just how they will do their voting and there isn't much discussion about those things anymore. Rather we are beginning to plan together that regardless of how the election will go, we will be better prepared to sooner work for the things that will be of benefit to the majority of the citizens. Sometimes I think business people and farm folks are seemingly beginning to regard politicians and political affairs as something of a necessary evil.

The long trip, away on one road and returning on another, with the corn picker gave me an opportunity to closely study farms I passed and I do not find the improvement I hoped for. Prices are better, but the reduction programs and the drought have not left very much to be sold at the better prices and I am just afraid things are not as much better as some folks think and hope for.

Fri. Oct. 30. 36.

Last winter I thot I had completed all of the land business and settled in full with the old mortgage company at the time I arranged the Federal Land Bank Loan, but yesterday along comes a letter from the Life Insurance Company about taxes and other matters. I opened the safe and looked thru the files and found a letter from the Insurance Company saying that everything was closed up. However, I wrote to them to have them clear their files completely. This afternoon I left off work and went to the court house and found that the old mortgage had been released.

While I was in the court house I inquired about the progress of the soil conservation work and found that there have been some complaints. During the old corn-hog program there was some misunderstanding on the part of some of the farmers. So when we went into the soil conservation program I paid closer attention and I am sure that there have been some misstatements made in the progress of this program from last spring until this fall. My own personal soil conservation business has been entirely satisfactory so far.

Sat. Oct. 31. 36.

The weather was more damp today and we used the picker about all day. We moved the portable elevator to another crib today. I settled up with the hired man and am paying him five cents for the corn he husked and $2.00 for the day work. I will not have him anymore at these day wages for two reasons, it is too much for me to pay and I can get cheaper help.

We went to town tonight and I had a broken casting welded for the picker. Quite a few folks were in town. At the theaters I found that there is beginning to be a reaction to the Bank Night policy. At first a great many people were hopeful that they might win the sum of money, but as they are realizing that that can happen to only one of the audience, they are turning to other and cheaper theaters.

Mon. Nov. 2. 36.

Tomorrow we must vote at the Presidential Election and we must go to rather an inconvenient place to vote. We did not listen to any political talks on the radio and did not read much in the paper. When the election is over, however it goes, the common average citizen must continue at his humble job and pay the bill, if it is ever paid.

Tue. Nov. 3. 36.

Today is election day. And when we had finished the morning work we drove to the school house where our township votes and cast our ballots. We are much more interested in the county offices than any of the others.

I can recall several elections and I do not think I can remember of another one where more money was spent both directly and indirectly on an election than on this one. Late tonight or sometime tomorrow we will know the results.[14] I think everyone has their plans pretty well made up as to the conduct of their business in the future. It will be just a little easier, or a little harder, as we look at it and perhaps plenty hard for most of us either way.

We husked some corn again today. The fields are wet and heavy and I doubt if we can use the picker anymore this season. Corn husking is about over. Many wagons have the throw board off again, which indicates that husking on that farm is finished. And the cribs are pitifully empty.

Fri. Nov. 6. 36.

Our two corn wagons continue in the fields. We have finished all but the last field and began on that one today. Our cribs are going to be very empty, and they look like spring already. A farmer friend remarked the other day that cussing the empty cribs, the drought, the President, the soil conservation scheme that did not make any provision for a drought shortage, all of these would not help now and he wasn't certain that continued blundering in the future would help either. Also he did not say just what he thot was the solution of a coming very serious problem.

Mon. Nov. 9. 36.

This morning there was enough of the Saturday snow covering the ground to leave it with a white blanket, but by noon it was mostly all gone.

This afternoon we were husking corn again. The fields wet and muddy and walking down the corn rows was tiring work. Gobs of mud were clinging to our shoes and we were continually slipping to the center of the space between the corn rows. The corn we were husking in was some better than the other parts of the field and our loads became heavier, causing more work for the teams. The sky was clear and the air nice tho. This evening I am planning the work so that we will finish the husking tomorrow. If one could be sure that the weather would be good for husking there wouldn't be any reason for hurrying the husking because this crop doesn't warrant spending any more money on it.

Tue. Nov. 10. 36.

We finished the husking today. Dark, cloudy weather this forenoon did not look favorable, but this afternoon was better. I am much disappointed in our corn yield. Of course I had some time to prepare myself for it. And as many of the neighbors have finished their husking I could see what their yield was. However, I had still hoped for a better crop than this was. As near as I can tell it will be between 12 and 15 bushels per acre. I do not know just now what I will be able to find to do, or how to make up this shortage. Another letter came today from the County Treasurer about my delinquent taxes. I have been planning to use my soil conservation money to pay the taxes and I think there will be enough of it for this when it comes. Then there is insurance too.

Wed. Nov. 11. 36.

One week ago yesterday was the election and already there is much talk and speculation about the future. Our county is planning to extend the city water mains eight or nine miles out in the country to the poor farm.[15] This is not regarded by anyone as a necessary or worthwhile thing to do. It is supposed to create employment and the Government will pay the bill.

Thur. Nov. 12. 36.

I spent the forenoon around the place. Our township soil conservation committee man was here and I signed the last papers. I have a total of 21 and a fraction acres for soil conservation and will be paid on 18 and a fraction. In our township we receive $13.83 for soil conservation and $1.00 per acre for grass seed. $2.00 for alfalfa seed, but I did not happen to use any of that. I should receive around $280.00 for my part in the program. 90% should be in the first check and the other ten percent will come later, with perhaps a small deduction for possible administration expense that will not be cared for by other funds. I do not suppose that I will be fortunate enough to receive this soil conservation money before I must make some other arrangement about paying the taxes. If I would have had a reasonable corn crop and been able to sell some of it for a reasonable price I could pay the taxes and the Land Bank payments quite easily. I think there will be corn to spare to pay the Land Bank, or I will see that there is enough left from the feeding, cut down the feeding enough in other words. And if the S.C. money will come in for the taxes we will get thru, but there won't be any corn or anything else to sell for any other things, the things that we are supposed to buy so labor will be able to buy ours or imported products. It is pretty hard to be classed along with some of the foreign farmers and to see their products in town along with ours.

Sun. Nov. 15. 36.

Fine weather again today. I was much tempted to make a long drive today. I am very partial to driving around the country to see how farm folks are getting along. But I did not go anyplace today. And I was glad that I did not, because city company drove in and I would have missed seeing them. They report everything in their town seemingly going along well. Everyone happy and contented. Some complaint about high food prices and there is beginning to be a slight slacking up of factory work. Just shorter days by an hour or two. And rather more frequent day's closings for most any reason.

Naturally food would be high. It is becoming scarce in some instances and in others there is no doubt about it. Almost profiteering is indulged in by some dealers in some lines. Country merchants are offering 34 cents for eggs and poultry prices are continually becoming lower. As feed costs mount, stock and poultry will go to market, then finally there will be a shortage of those things and imports will result.

Tue. Nov. 17. 36.

Tonight we drove to town to see a Will Rogers' show and enjoyed it very much. Since his death I often wish he could have had the leading part in several farm films. He could do the part of a farmer very nicely.[16]

Wed. Nov. 18. 36.

My new gasoline lantern is repaired and L.L. drove to town late this afternoon to get it for me. It is very nice to have it to use again. Often I wish for light and power from the high line. However, we frequently hear of farm folks who are having their lights disconnected because of the difficulty of meeting the bills. A farm usually can turn out only so much money and when the weather or the markets (in this case the weather) does not provide enough revenue one of two things may be done. Borrow money or do without. Doing without seems to be the safer way for some folks.

Sun. Nov. 22. 36.

Another very fine Sabbath Day has passed. I did not attend services anyplace. I drove over into Greene county for a short time this morning. Talking over some of the farm problems among we farmers may result in a few of us being able to avoid some of the difficulties ahead of us. Many farmers believe farm organizations should be strengthened but they freely say they do not see how they could pay a $10.00 membership fee to belong to the Farm Bureau. Some farmers say this will be the end of this one. Other farmers say that the New Deal cannot entirely control this one and are taking this means to wear it out and then they will bring out a new farm organization.

Mon. Nov. 23. 36.

Here and there I see a farm that is almost entirely newly fenced. Inquiry usually reveals the fact that it is a farm owned by a city man. Usually in the grain or lumber business. Two of the best improved farms in our neighborhood are owned by city men. One, an out of the state man, has a flour mill. The other owner is in the grain business. Both of them have been able to do very well the past several months, mainly by speculating. These farm owners are not popularly mentioned in our neighborhood, tho their tenants are neighbored with the same as all of the farmers in the community.

Tue. Nov. 24. 36.

I drove to the county seat this morning. All summer I had planned to use the soil conservation grant money to pay the taxes with. Now I find that the farm has to be advertised in the tax sale because I have not paid the taxes. I was finally able to make arrangements so that it is supposed to not be put up for sale at this time. If I would have had the usual corn crop so I could sell corn to pay taxes or if I would have had the soil conservation field in a crop that I could sell now I would have had the tax money. If I sell the present small pile of corn to pay these taxes I will not have enough to feed and make the Land Bank payment. Last year the farm was sold for taxes two days before the corn loans came out. And still they expect the farmers to continue to put up with this kind of blundering and bungling. I am hopeful that my soil grant will come in time. But quite a few papers are coming back for various reasons. I have tried to be very particular about having mine exactly right.

Thur. Nov. 26. 36.

Today is Thanksgiving Day. I did odd jobs around the farm and attended the regular Thursday sale at the sale barn in Jefferson this afternoon. The economic situation is becoming so acute that many farmers think they must continue right on the job all of the time. The offering and attendance were about the same as usual, but values were some lower. There is quite a bit of talk about the small numbers of live stock on the farms, but the feed supplies are equally or more reduced.

The division between the two money classes each day becomes more and more apparent. Business men, professional men and teachers and other salaried people and the farmers who happen to live in the section where they had a good corn crop make up the money folks. Laborers, those on relief, farmers without any or much crop etc. make up the no money class.

Fri. Nov. 27. 36.

Very nice weather again today. We were busy with repairing fences all forenoon. Using all of the old material that we possibly can, it is taking more time than I had planned for fencing.

This afternoon we drove to the county seat and I spent two hours at the community sale. Male hogs sold higher today than they did yesterday. The difference in the location of the sale barns accounts for this. There are parts of our county that produced a fair corn crop and it is adjacent to territory that received more moisture last crop season. Therefore buyers have more money to spend. For some little time I have noticed that some things around the sale barn were not conducted quite as they should be and today several farmers came to me to talk about how legal rulings were being overlooked to the advantage of the speculators and the loss of the producer stockmen.

Here, as in the sale yesterday, in the audience of buyers and sellers I could notice those who are in the money and those who are not. Perhaps the location of our county, rather between or on the edge of the drought area, makes this more noticeable. Farmers continue to remark about the movement of corn from farms where it seems that it will surely be needed before another crop is produced.

Sat. Nov. 28. 36.

The weather was fine again today and we made good use of it. Repairing fences, hauling and sawing pole wood and grinding feed. We also moved poultry houses.

The very low prices of poultry have been very discouraging to many producers. Many of who still feel that there has not been a large production to cause the lower prices but that it is market manipulation, and that later in the season after the poultry has left the farms a shortage will be discovered and prices advanced to the great benefit of the storage men. Numerous instances have come to my attention where grades have been changed and prices juggled and the producers, when the sale or pool has been completed, find the receipts very much lower than they were led to believe they would be. A letter came today from dependable sources, giving information where reliable Hy-Bred seed corn may be found.

We were in town a short time tonight. Our cream hauler missed us the last time and tonight I took our cream to another firm and found that the test was just exactly double the amount our last test was.

Sun. Nov. 29. 36.

The fine weather continues. I visited a short time this morning and this afternoon drove to see an old neighbor who owns a farm in our community. A land speculator, who owns a farm near here, is attempting to have a gravel road constructed that will benefit only his own farm but would be taxed to other adjoining farms whose owners are, or some of them at least, heavily burdened with taxes. I found this old neighbor very much opposed to any additional taxes of any kind until much of our present expenses are paid for.

Tue. Dec. 1. 36.

Today is the beginning of the last month of this year. While there are many things around the farms that should be done "right now" whether we like it or not, we must pause sometimes and take a moment to think about the closing of this year and the circumstances that have been instrumental in shaping the events of the past year. Because of changing conditions affecting the future, I am afraid there isn't much in the past year that one will be able to use in planning for the coming season. The year of 1937, as [far as] the farming business is concerned, is a new and uncharted sea to be sailed almost blindly.

Thur. Dec. 3. 36.

Recently I have been having trouble with a small leak in the car radiator and some of the anti-freeze was being lost. This evening I drove to town hoping to have a permanent repair made but was too late to have the work done and attended the movie "Tarzan Escapes." I was not so much interested in Tarzan as I was in part of the picture showing the animals and the lay of the land in that part of the country. I wondered if those wild animals would give way to domestic ones and the country ever become a farming country.

Mon. Dec. 7. 36.

Since last Friday D.L. has planned to market turkeys thru the pool today. Yesterday he had rather given up the idea for a day or two because of the stormy weather, but today he decided to go ahead with it. And I thot too that would be the best thing to do. So this morning I drove a 16 mile trip for the coops and several of the neighbors assisted and we cooped the birds that were to go. I trucked them about 50 miles and arrived at the Coon Rapids plant of Armour and Company just after noon. D.L drove in his car and remained until after the dressing and grading was completed. It was a cold drive for me, but it was in a country I was not familiar with and I noticed many things of interest to me as I was passing. Mainly the absence of livestock on the farms. The weather was quite cold and not much of the stock in the fields, but it could be seen in the lots near the barns. I think the drought was quite severe here, but it has laid a more heavy hand on that part of the country.

Many turkey growers were at the dressing plant today and they all seemed to feel the same about the situation. Some of them will continue next year and those that do plan to continue are having trouble to find male birds for next season's matings. It is said that Armour and Co. will not buy turkeys but handle them on a percentage basis. In fact I do not know where a buyer could be found.

Fri. Dec. 11. 36.

Quite a busy day again today. After doing the morning work I put the stock rack on the truck and loaded two shoats and a calf and drove to Bill's place where we loaded a male hog and a calf for him and then drove to the community sale. Where I left the calf and Bill [left] his calf and hog. From there I trucked my hogs to the Swift buyer and received $9.00 per cwt. for 170 lb. hogs.

Returning to the sale barn for lunch and the beginning of the sale, other members of the family having arrived with the mail they had picked up when passing our box, I learned that my soil conservation check had arrived. Driving to the court house I received a check for $250.70 as the first payment. There should be a 10% check to come later. As soon as I had the check I went at once to the office of the County Treasurer where I paid my taxes. The delinquent tax sale was held last Monday, but the Treasurer had not listed our place, knowing the soil conservation money should be in soon. This is a favor I appreciate very much. I paid $198.00 real estate, $7.50 personal and $11.40 gravel road tax.

Wed. Dec. 16. 36.

Bill and I went in town tonight. We went to a movie and while we were there my name was called across the street in another theater where they were having "bank night" tonight. The amount was $300.00. I have only attended this bank nite theater once in the last month and I did not register then. In fact I cannot recall that I ever did register in any theater. I think it is an attempt to encourage farm attendance. I suppose I will hear a lot about it for the next week or so.

Thur. Dec. 17. 36.

Off and on I have been trying to use a microscope in studying the seed germs of the various farm seeds that I use. I have never been able to find one useful in determining if there was life in the seeds. Recently I tried another scope, thinking there might be improvements in them. This last one I think does show the difference between the live and the dead seeds, but it takes much more time with the scope than to test the seeds for natural germination.

Fri. Dec. 18. 36.

Nice weather today and a good day to work around the farm, which I did this forenoon. However, this afternoon I went to town to see about having a drainage culvert placed thru the roadway adjoining the farm.

If I drive east a county or two I find much evidence of prosperity and if I drive west an equal distance I find evidence of the other condition. Much of it is noticed by the cars parked in town. They tell a very plain story. And of course there are some signs right here at our place that indicate that the drought came uncomfortably close to us.

Sun. Dec. 20. 36.

The year that is drawing to a close is mentioned more frequently, by more folks, than any year I can remember of. The happenings of 1936 must have made a deep and lasting impression on many people and these happenings are still close in their minds.

Tue. Dec. 22. 36.

For some little time a small quantity of corn said to be from South America has been here on my desk. Occasionally I look at it and turn it over with my pen or my pencil. I am still worrying in my mind about a situation like this. I am absolutely certain that I would be in better shape financially if I would have produced or attempted to produce an average crop and received an average price for it. And I am not at all certain that the methods we are following will ever lead to the proper solution of the production, price problem.

Wed. Dec. 23. 36.

I have a farmer friend, who lives in the territory where sufficient rain fell last summer to make him a good corn crop. His yield was near 40 bushels per acre. With the present price he is quite prosperous. He has a new car, a new tractor, a new two row corn picker and now he is shaving with a $15.00 electric razor. In another community I met a business man who said he had $700.00 on his books and that he drove among the farmers for one day and did not succeed in collecting one dollar of this amount.

I am just a little worried about my prosperous farmer friend. I can recall other years and other times when other farmer friends were just as prosperous as this one above mentioned is and now they are out of the picture altogether.

I sold a few hogs today, received nine cents for them. Applied the money on a note at the bank. Helped D.L. dress more turkeys today.

Fri. Dec. 25. 36.

Today is Christmas Day. The weather was fine all day and evening. Mother desired that as many of the family as possible spend the day and have dinner with her. Not all were able to do so. But the greater number were there. Our family did not exchange as many gifts this season as usual. We just did not seem to do it, that is all. A New Year's Day dinner is planned for next week, but in many ways today seemed more like New Year's Day than Christmas. I suppose it was because we spent more time talking about the happening of the past year and of the coming one than we did about Christmas. Several outstanding tragedies of the past year will remain long in our memories. We are a farming family and the drought and its effects on us and other farm folks will long be remembered by us. Several older members of the family seemed to suggest they were not pleased with some of the governmental activities as regards agriculture. All members of the family regretted and deplored the increasing numbers of highway accidents and deaths. Our family have been very fortunate in this respect. But we feel that it is one of our greatest impending dangers.

Sat. Dec. 26. 36.

This afternoon D.L. drove to an adjoining county seat and I accompanied him. For some little time I have been hearing about a small community sale

barn, away in one side of the town, and I thot this would be an opportunity to
visit it. There wasn't any livestock in this sale. All of the various articles that
one could think of from country and town homes. I spent some time looking
over the offering. And I spent more time looking over and talking with the folks
who had brot in the offering and the possible purchasers. It was a very unpleasant afternoon in all respects. People were parting with anything they
possessed that might bring in a little cash for immediate needs. Hundreds of
empty glass fruit jars were offered. Many of them just recently emptied and
should have been kept for next season. But we must have a little money was a
statement frequently heard. And a very little money was all anything brot.
One learns many things from these people after you learn to understand them
and have their confidence. One man said he had rented the same farm for more
than thirty years, had paid over $30,000.00 rent on it and just now when the
children were grown and educated and he thot he was going to be able to buy a
small farm, the drought was going to cause him to lose everything. He complained bitterly about the economic system and I had to agree that in his
case the Government and all others had acted exactly wrong. Others, not half
so deserving, were receiving not nearly so much needed assistance.

Sun. Dec. 27. 36.

We spent the day quietly at home and attended services in the village
church this evening. The last Sabbath Day of the year is passing. Thinking
over the past years it seems to me that the next few years will see the passing of
many rural and village churches unless they are able to some way consolidate
their congregations and efforts. Too many rural and village churches have
already discontinued. And the city is not the place for farm folks to worship.
Rather city folks should worship in the peace and quiet of the country with the
country folks.

Mon. Dec. 28. 36.

It seems that always I mention the weather first. And it is to a very great
extent perhaps one of the greatest controlling factors of farm life. Damp, misty
weather today made it unpleasant outdoors. With colder weather a film of ice
formed on things and this made travel for those who were out quite dangerous.

Tue. Dec. 29. 36.

At the court house I heard some talk about wage increases for all highway
employees, with a minimum of 80 cents per hour. I walked several blocks away
from the court house looking at the parked cars and saw only three with new
number plates. They were not farmers cars. I happen to know that the average farmer's income per hour in that particular county is nearer eight
cents than 80 cents per hour. But the farmers are good sports and will continue
to try to feed the nation and perhaps pay for the privilege of doing it.

Thur. Dec. 31. 36.

Hog marketing, for many farmers, was the business event of the day. Our local yards were overcrowded with hogs. Prices above ten cents were paid for nearly all that were bot. Talk around the yards indicate eleven to twelve cents will be paid during the next few weeks. Not many seem inclined to hold for that price. Shortage of feed, low stock of corn etc. is the reason for selling at this time. Several farmers said they had kept them too long now and cut into stock feed they should have for next spring's work.

I have written 1936 for the last time and tomorrow a New Year begins. I am facing it with a full realization of some of the things that may be ahead for farm folks and I hope there will not be too many disappointments and discouragements for us and them.

ELMER POWERS'S STRUGGLE did not end in 1936. He continued through the next years to contend with low prices and the complexities of the federal farm program. Yet the difficulties were probably never again quite so acute as they had been during the 1934-1936 drought era. Iowa's fickle weather moderated, the second farm program seemed to run more smoothly than had the first emergency measures of the New Deal, and the coming of World War II promised recovery.

The farm operation slowly came to be more mechanized as the 1930s drew to a close. Elmer invested in new and more powerful equipment, and in 1939 a long-term wish was fulfilled when an electrical power line reached the farmstead. He, typically, had been involved in setting up the local REA program—drawing line maps and attending committee meetings. However, a utility line from Boone reached the farm before the REA installation was complete. As electric and gasoline powered equipment was installed, the number of horses on the farm dwindled until by the early 1940s only one or two remained for light work.

The change was timely since the offspring of the Powers family broke away from home in 1939. D.L., who had for some time been occupied away from the farm with his feed and locker businesses, married in 1939 and moved into a nearby town, beginning a career in agribusiness that would make him prosperous within a few years. L.L. also married the same year and, after a brief sojourn in Illinois, returned to live with her husband on a farm near Amaqua Township.

On the home place, Elmer still operated on a slim margin. He continued to struggle with loans and federal payments to keep his farm afloat. Prices held fairly steady and crop-support payments stabilized, but his cash income remained low. In March 1941 Elmer recorded that his taxable income was a mere $741 (too low to require a tax payment), and he noted, "As far as I am able to learn this is a larger net income than many of my neighbors have."[1] The cumulative effect of the Depression is better understood when this figure is compared to the $2,000 annual income figure he had recorded during the mid-1920s. Elmer neatly summed up the problem: "Too much reduction program, hard times, and drought."[2] Nonetheless, the worst crisis seemed past, although the mild renewal of the farm economy brought problems that bothered him. His diary entries show a concern for the rising rate of farm accidents and fires, which he attributed to increased use of power machinery and carelessness among farm people. The Depression struggle had taken its toll. In March 1937 he wrote of the death of a young farm woman: "There are several factors contributing to this run of sickness and accidents and the principal one is the hard times that have been with us so long and are still with us just about as bad and in some cases worse than ever. In the case of the farm lady, she struggled along for the past few years, everything getting from bad to worse until a complete breakdown was the result."[3]

Elmer's comment was prophetic. In the same year he began to note in his "private" diary frequent trips to the doctor in Jefferson. His health became worse as time and hard work bore down on him. The addition to his family of a grandchild—D.L.'s son Dennis—brightened his days, but by the early 1940s it was clear that Elmer was seriously ill.

The beginning of the war in Europe and America's eventual involvement

sparked the long-awaited recovery for agriculture, but Elmer Powers was to share in it only briefly. He operated his farm with the help of a procession of hired men, struggling with new wartime problems of fuel, tire, and machinery shortages. Until the end of his life, despite prolonged pain from his illness, he continued to do the heavy farm work. His last diary entry reads: "Some snow this noon. Stopped corn husking for the day. I have my A gas book but do not have any truck papers. Also the tractor is still an orphan as far as fuel is concerned."[4]

Elmer Gilbert Powers died on the day after Christmas 1942 at the age of fifty-six. The official cause was kidney failure but, like his father before him, he had fought an unsuccessful battle with cancer.[5]

The quarter-section farm that had encompassed his life remained in his family, passing eventually to D.L. Today it is part of the Powers Land Company holdings. Elmer's grandson, Dennis Powers, lives in the old farmhouse with his family, including a young son who is the sixth generation of Powers to walk the flat, rich fields.

"QUIETDALE FARM," 1974. LOOKING FROM SOUTH TO NORTH.
(L. Edward Purcell Photo)

INTRODUCTION

1. The four sons of Samuel (1855-1933) and Kate Powers (1862-1950) were Samuel A. (1884-1974), Elmer G. (1886-1942), Walter W. (1889-1959), and Daniel I. (1895-1974). The family also included two daughters, Elizabeth S. (1891-1944) and Elnora, who was adopted. Their first-born son Albert died in infancy.

   For a biographical sketch of Samuel Powers see N. E. Goldthwaite, ed., **History of Boone County** (Chicago: Pioneer Publ. Co., 1914), pp. 371-72. See also the **Ogden Reporter,** Nov. 9, 1933.

2. Goldthwaite, **History of Boone County,** p. 372.

3. Personal interview with Daniel L. Powers, June 19, 1974. A transcript of the taped interview is in the Powers Papers of the State Historical Society of Iowa, Iowa City, hereafter referred to as **DLP Interview.**

4. Donald R. Murphy to Dr. William J. Petersen, Aug. 27, 1953, Records of the State Historical Society of Iowa, Iowa City.

5. **DLP Interview.**

6. Elmer G. Powers Diary, Apr. 9, 1931. This document is in the Powers Papers of the State Historical Society of Iowa, Iowa City, hereafter referred to as **EGP Diary.**

7. **EGP Diary,** Nov. 8, 1931.

8. E. G. Powers to Henry A. Wallace, Nov. 9, 1932, Henry A. Wallace Papers, The University of Iowa Libraries, Iowa City.

9. E. G. Powers to Henry A. Wallace, Apr. 29, 1933, Office of the Secretary of Agriculture, General Correspondence, The National Archives, Washington, D.C., Record Group 16.

10. **EGP Diary,** May 29, 1938.

11. Murphy to Petersen, Aug. 27, 1953.

12. Between 1931 and 1949, **Wallaces' Farmer** published short extracts, usually only a sentence for each day, from Elmer's diary. Most of the extracts were fairly mundane, dealing with weather or prices, for Editor Murphy cut Elmer's more provocative comment on politics or the social system. The diary was used primarily to remind readers what conditions had been for the farmer during earlier stages of the Depression. Several of these previously published extracts are included in this edition of the diary and are reproduced through the courtesy and permission of **Wallaces Farmer.** Following are complete citations to published extracts: **Wallaces' Farmer**

and Iowa Homestead, 56(Nov. 14, 1931):18; 58(Jan. 21, 1933):38; 61(Jan. 4, 1936):13; 61(June 6, 1936):11; 63(Jan. 29, 1938):5,14; 63(June 4, 1938):8; 63(Sept. 10, 1938):16; 64(July 1, 1939):15; 65(May 4, 1940):3; 65(Nov. 16, 1940):10; 74(Mar. 5, 1949):312.

CHAPTER 1: DAY BY DAY ON THE FARM: 1931-1932

1. The village is Beaver, Iowa, two miles southwest of the Powers's farm home. The county seat is Boone, Iowa, thirteen miles east.
2. D. L. is Elmer's son Daniel L., eighteen years old when the diary began.
3. L. L. is Lillian Lenore, Elmer's daughter and not quite thirteen years of age.
4. Elmer seldom purchased seed corn or hybrid corn. He selected instead the most cosmetic ears from his own harvest and stored them over the winter to provide the next season's seed. The germination qualities of such seed were unknown until planted.
5. This number seems high but, relying on horses, individual farmers could not manage large acreages. A quarter section was considered an average and adequate size of farm in Elmer's neighborhood. Thus there were many more farmsteads in this vicinity in 1931 than today.
6. This was undoubtedly Greene County, directly west of the Powers's home. Elmer frequently visited communities there—Jefferson (the county seat), Dana, Grand Junction, Paton, and Rippey.
7. Brother Bill is a pseudonym for Elmer's brother Walter.
8. The town is Ogden, Iowa, six miles east of Elmer's farm.
9. Des Moines, Iowa, about forty miles southeast of the farm.
10. Elmer refers to the editors of **Wallaces' Farmer and Iowa Homestead**, the premier agricultural periodical in the state and one of the leading farm publications in the nation. The diary was being written for Donald Murphy, associate editor. Henry A. Wallace, soon to be appointed Franklin D. Roosevelt's first Secretary of Agriculture, served as editor.
11. The Powers brothers banded together to form their own threshing "ring" at the urging of their father. It was a common practice for several farmers in a neighborhood to form such cooperative groups, sharing the expense of operating the thresher and the work time necessary to complete the threshing for everyone in the "ring." Feeling that others in the community were short-changing them, the Powers family pooled resources in the early 1900s and purchased a Wood Brothers threshing machine. Each brother donated time until the work for all was done. Occasionally, the family did custom threshing for a few outsiders. **DLP Interview.**
12. Iowa State College (now Iowa State University) in Ames.
13. The annual Iowa State Fair was (and still is) a great event in the summer schedules of many Iowans, especially rural families. Livestock and homemaker shows, horse racing and entertainment, horse-pulling contests and butter sculpture all went to make this an exciting and pleasant break in farm life, falling conveniently between the small-grain harvest and corn-husking time. Originally a farm show, by the 1930s the State Fair

was already well on the way to becoming the show business extravaganza that it is today.

14. Iowa's celebrated "Cow War" was touched off by a dispute over enforcement of the state's bovine tuberculosis test law. Farmers felt the regulations, which called for herd testing, were costly and unnecessary, despite evidence that numerous cases of T.B. in children could be traced directly to milk from infected cows. Governor Dan Turner stood behind the law and pushed for strict enforcement. Trouble had brewed since March 1931 and finally bubbled over on September 22 when farmers near Tipton violently resisted a renewal of testing. Troops were called in, and eventually calm was restored. The incident illustrated the raw nerves brought on by hard times. Earle D. Ross, **Iowa Agriculture** (Iowa City: State Hist. Soc. Iowa, 1951), pp. 164-65; **Des Moines Register,** Sept. 22-23, 1931; Theodore Salutos and John D. Hicks, **Twentieth Century Populism: Agricultural Discontent in the Middle West, 1900-1939** (Madison: Univ. Wis. Press, 1951), pp. 437-41; Roland A. White, **Milo Reno, Farmers Union Pioneer** (Iowa City: Iowa Farmers Union, 1941), pp. 49-64; Frank Dileva, "Frantic Farmers Fight Law," **Ann. Iowa,** Ser. 3, 32(Oct. 1953):88; John L. Shover, **Cornbelt Rebellion** (Urbana: Univ. Ill. Press, 1965), pp. 28-33; Walter Davenport, "Get Away from Those Cows," **Collier's** 84(Feb. 27, 1932):10-11.

15. Elmer, like most of his neighbors, did not use a power-driven corn picker during the early 1930s, although he had owned one since World War I. Corn husking usually began in October and continued into the winter until all the crop was harvested. Huskers walked the corn rows in company with a horse-drawn wagon. The ears of corn were stripped from the stalks by hand, and the husks were twisted off with the help of a peg or hook. This was dreadfully taxing labor. A particularly good husker could harvest 100 bushels in a day. **DLP Interview.**

16. A hired hand.

17. Al is Alvin Harten, a neighbor boy.

18. March 1 was the traditional moving day for tenant farmers in Iowa.

19. Much of the land in central Iowa was still close to virgin soil with reasonable capacity to supply nitrogen for corn. However, part of the nitrogen supply necessary for good corn production was tied up by decomposition of fresh cornstalks. In addition, plowing through a field of stalks was difficult if not impossible using only horsepower. This led farmers to burn cornstalks from the previous year in the fields. To handle the tough stalks, farmers broke them loose with a long pole—sometimes a railroad rail or a telephone pole—which was dragged through the field between two teams. After breaking, the stalks usually were raked into windrows and burned. **DLP Interview.**

20. Pat is a pseudonym for Dwight Gleim, the son of a neighbor. Gleim often helped on the Powers farm during the early 1930s. Elmer G. Powers private diary, Jan. 1, 1933. Elmer wrote a "private" diary account, parallel to the one that he kept for **Wallaces' Farmer.** A copy of the private diary is in the Powers Papers of the State Historical Society of Iowa, Iowa City, hereafter referred to as **EGP Private Diary.**

21. Although it was 1939 before an electric transmission line reached the

Powers farm, Elmer purchased a small gasoline-powered electric plant from the village of Beaver shortly before World War I. According to Dan Powers, this plant "broke frequently and was a constant headache to operate." **DLP Interview.**

22. Bobbie is Robert Powers, a son of Elmer's brother Walt.

23. The Farmers' Holiday Movement was the most important midwestern agrarian protest during the Depression. The Movement centered around the fiery Iowa radical, Milo Reno, and the Farmers' Holiday Association, an outgrowth of the older Farmers Union. Reno led his first farm strike in 1920 and continued for the next two decades to agitate against government policies and low prices. In 1932, as support grew among farmers for direct action, Reno and the Holiday Association called for a major farm strike on August 8. Strikers were determined to withhold produce from the marketplace as a way to force prices up and, more importantly, to pressure lawmakers for permanent relief. Reno's manifesto was clear: "Merchants, Manufacturers, Newspaper Publishers and Union Labor refuse to deliver their products until the price they set upon their salable services and wares have been met. We insist upon the same right to conserve our resources and demand the same fair prices as do other industries."

    The protest was strongest among milk producers in the Sioux City, Iowa, area, although support was widely scattered throughout the Hawkeye state and parts of the Dakotas, Minnesota, and Nebraska. One source counted 1,500 strike adherents in Elmer's own Boone County. In addition to withholding their produce from the market, the strikers set up picket lines to prevent nonstrikers from making deliveries. Through much of August, violence flared between striking farmers, nonstrikers, and sheriffs' deputies. The most complete study of Reno and the Holiday Movement is Shover, **Cornbelt Rebellion;** White, **Milo Reno** gives a helpful but partisan view; and Everett B. Louma, **The Farmer Takes a Holiday** (New York: Exposition Press, 1967) is sketchy. The statewide impact of the Movement can best be gauged from daily accounts in publications such as the **Des Moines Register** and **Wallaces' Farmer,** and the Holiday was accorded much publicity and comment in the national press and magazines at the time.

    Elmer Powers showed little sympathy with the Holiday Movement and never joined the strike. In fact, Elmer, in a letter to Secretary of Agriculture Henry A. Wallace of Apr. 29, 1933, said, "From what I knew of the Farmers Union I always tried to keep clear of it and when the Farmers Holiday Association started I avoided those folks as far as taking any part in things was concerned. . . ." Office of the Secretary of Agriculture, General Correspondence, The National Archives, Washington, D.C., Record Group 16.

24. On October 4, 1932, President Herbert Hoover opened his reelection campaign by speaking in his home state to an overflow crowd at the Des Moines Coliseum. In his address Hoover outlined a twelve-point program designed to revive American agriculture. For example, he called for the maintenance of high protective tariffs on farm products; relief to farm owners from pressing long-term mortgage payments; and extension of loans,

through the Reconstruction Finance Corporation, to processors of agricultural products. **New York Times,** Oct. 5, 1932.

25. The Democrats swept normally Republican Iowa on election day 1932. Roosevelt won by 183,586 votes, and the Democrats captured the governor's chair and all other state offices. In Amaqua Township FDR won by 96 to 21. The totals for all of Boone County were 5,293 for Roosevelt to 3,694 for Herbert Hoover. **Iowa Official Register, 1933-34,** p. 226.
26. The Bill referred to here is a temporary hired hand, not Elmer's brother Walt whose pseudonym was also Bill.
27. This reference is to a neighbor, Jim Peachey, not a previous hired hand with the same first name. **EGP Private Diary,** Nov. 13, 1933.

CHAPTER 2: COMING OF THE NEW DEAL: 1933-1934

1. The shipping association consisted of local people who pooled their hogs to make carload lots. They then shipped the cars of hogs to a terminal stockyard where a commission man sold them to the packer. The purpose of the association was to retain the local hog buyer's profit for the farmer. This particular shipping association lasted for only a brief period. Letter from D. L. Powers to authors, Mar. 11, 1975.
2. Clyde Herring, Iowa's new Democratic governor, issued a proclamation on Jan. 19, 1933, which asked financial institutions to cease foreclosures on Iowa farms. Directed largely at life-insurance firms that held more than 40 percent of the state's farm mortgages, the request was agreed to by most companies. By Feb. 1, 1933, Aetna, Connecticut General, Connecticut Mutual, New York Life, Phoenix, and Prudential announced that they would suspend foreclosures in the Hawkeye state until legislation improved the overall status of debtors. **New York Times,** Jan. 31, Feb. 1, 1933.
3. President-elect Franklin D. Roosevelt announced the selection of Henry A. Wallace of Iowa to be his Secretary of Agriculture on February 12, 1933. Wallace came from a Republican family—his father Henry C. Wallace had been Secretary of Agriculture under Warren G. Harding. The younger Wallace, however, split with the Grand Old Party over farm policy. He became an ardent supporter of active government controls to limit surpluses and to correct the international trade imbalance, which he felt caused falling commodity prices. He had vigorously supported FDR in his publication and was eager in 1933 to move ahead with relief and reconstruction for the farmer. Wallace was a prominent farm editor, corn plant biologist, and statistician when he left for Washington at age forty-four. A standard work on Wallace for this period is Edward L. and Frederick H. Schapsmeier, **Henry A. Wallace of Iowa: The Agrarian Years, 1910-1940** (Ames: Iowa State Univ. Press, 1968); see also **Wallaces' Farmer 58**(Mar. 4, 1933):5.
4. The nation's finances reached rock bottom during the interregnum between Hoover and FDR. The banking structure literally collapsed, with many banks closing their doors as panicked depositors lined up to demand payment. At one o'clock on the morning of March 6, 1933, President

Roosevelt issued a proclamation of a national bank holiday, temporarily closing financial institutions throughout the country. Congress quickly followed the holiday with passage of the Roosevelt-sponsored Emergency Banking Bill, a measure that made it possible for banks to reopen if they were found to be solvent. **The Public Papers and Addresses of Franklin D. Roosevelt, Vol. 2, The Years of Crisis, 1933** (New York: Random House, 1938), pp. 24-29.

5. Elmer's reference to the "change in the German Government" is somewhat misleading. The American press announced on March 21, 1933, that the Nazi government under Adolf Hitler would likely assume dictatorial powers; however, the Enabling Act to achieve that end did not pass the Reichstag until March 23. This measure gave Hitler direct powers to legislate without a parliament beginning April 1, 1933, for a four-year period. **New York Times,** Mar. 21-24, Apr. 1, 1933.

6. On March 13, 1933, President Roosevelt called on Congress to modify the Volstead Act and allow the manufacture and sale of beer and light wines. Elmer and other farmers resented the attention paid to the "beer bill" while farm relief measures waited.

7. The representative in the Iowa General Assembly from the Fifty-third District (Boone County) was Ben B. Doran of Grand Junction, a first-term Republican. **Iowa Official Register, 1933-34,** p. 73.

8. On April 27 and 28, 1933, the farmers' rebellion burst into frightening violence. On the first day, a mob of farmers in Primghar, Iowa, seat of O'Brien County, forceably halted a sheriff's sale on a local farm by overpowering 25 billy-wielding deputies. Later in the day, Judge Charles C. Bradley, senior judge of the Twenty-first District, was presiding at a foreclosure hearing in Le Mars, Iowa, when a lynch mob of disgruntled farmers broke into his court and carried him from the room. Outside, Bradley was partially disrobed, and a noose placed around his neck. His display of courage, however, calmed the mob, which dispersed after roughing him up. The following day, a mob of 800 farmers in Crawford County, Iowa, near Denison, descended on a chattel sale and beat up the sheriff and 50 deputies. Governor Clyde Herring declared martial law the same morning, and state militia moved into Le Mars and Denison. Following a show of force and the arrest of several farmers, the protest subsided. Nonetheless, the incidents stirred wide anxiety about an impending agrarian revolution in northwestern Iowa, and the affair received widespread publicity in the state and nation. For the best account of the disturbance see Shover, **Cornbelt Rebellion,** pp. 116-25.

9. The Red Ball salesman was a traveling vendor who offered oddments to farm families. Favorite items were candy and trinkets for the children.

10. The representative to the United States Congress from Iowa's Eighth District (which included Boone County) was Fred C. Gilchrist (not "Gilchrest" as Elmer spelled it) of Laurens. One of only three Iowa Republican candidates elected to Congress in the 1932 Democratic landslide, Gilchrist was in his second term. **Iowa Official Register, 1933-34,** p. 59.

11. After an initially optimistic, wait-and-see attitude toward the new administration, the Farmers Union and the Holiday Association became harsh critics of the New Deal, particularly when it ignored their currency inflation panaceas. The Agricultural Adjustment Administration (AAA) was Roosevelt's and Wallace's major program for agriculture. In general, the mammoth undertaking was aimed at curtailing production and at bringing about an immediate reduction of surpluses. Congress passed the bill during the Hundred Days period when FDR still commanded action from the lawmakers. The AAA was to have an immediate and significant effect on Elmer Powers and he referred to it frequently.

12. Codes were part of the National Recovery Administration (NRA), the federal agency created to administer another legislative product of the Hundred Days, the National Industrial Recovery Act. The purpose of the NRA, according to Roosevelt, was simply to put people back to work. The government established codes for cooperative action among trade groups and created a massive program of public works.

13. One of the first provisions of the AAA to be implemented was the purchase and slaughter of lightweight (immature) pigs and pregnant sows. A controversial emergency measure, the move was aimed at reducing the surplus of hogs on the market, thus driving up prices. The AAA arranged with packers to buy young hogs and "piggie" sows which would be processed into salt pork for distribution to needy people by the Federal Emergency Relief Administration (FERA). Between August 23 and October 1, nearly 6.2 million pigs and 22,000 sows were sold and slaughtered. Urban critics of the AAA howled in protest as immature pigs were killed before reaching full food potential. There was even a sentimental outcry against killing the "cute little pigs." See Van L. Perkins, Crisis in Agriculture: The Agricultural Adjustment Administration and the New Deal, 1933 (Berkeley: Univ. Calif. Press, 1969), pp. 134-46.

14. The corn-hog program was part of the New Deal's basic approach to agricultural reconstruction. The philosophy of the AAA under Wallace was to deal with the agricultural depression by reducing production. Corn and hog producers signed individual contracts with the Secretary of Agriculture, which specified that the farmer would reduce corn acreage and hog production. The Secretary then made cash payments to the farmer for both corn and hogs. Moreover, the government rented the idle corn acres, paying a fee based on the average yield for the preceding three years (later changed to a fee based on appraisal by a local committee). Two-thirds of the rent was paid to the farmer as soon as possible after he signed the contract, the rest in 1934 when evidence of compliance was shown. Farmers were also paid $5 per head for not marketing hogs (figured on 75 percent of the average number of hogs marketed during the previous two years). Of this price, $2 were paid immediately, and the balance deferred until compliance could be demonstrated. The program was run at the local level, with temporary committees conducting the sign-up. Elmer became closely involved in the early stages of this complicated administrative process. See Richard H. Roberts, "The Administration of the 1934 Corn-Hog

Program in Iowa," **Iowa J. Hist. Polit.** 33(Oct. 1935):307-75; Murray R. Benedict, **Farm Policies of the United States, 1790-1950** (New York: Twentieth Century Fund, 1953), pp. 302-15.

15. In 1933 the federal government embarked upon a massive program of public works. Under the FERA, headed by Iowan Harry Hopkins, the nation saw creation of the temporary Civil Works Administration (CWA) in November. By mid-January 1934, the zenith of the CWA, this agency had pumped a billion dollars of purchasing power into the depressed American economy. By spring 1934, however, the Roosevelt administration dismantled the CWA, fearing creation of a permanent class of "reliefers." The National Industrial Recovery Act (Title II) established the Public Works Administration (PWA) in June 1933. Led by "Honest Harold" Ickes, the Secretary of the Interior, the PWA proved more versatile than the CWA. It initiated its own projects, it made allotments to allow other federal agencies to carry on construction work, and it gave grants and loans to state and other government bodies to do nonfederal construction. Unlike the CWA, the PWA was not restricted to the use of labor from relief rolls. See A. W. MacMahon et al., **The Administration of Federal Work Relief** (Chicago: Public Admin. Serv., 1941).

Elmer was irritated, as were many farmers, by the creation of these New Deal relief agencies. He saw them as soft jobs with short hours. Elmer's criticisms of such programs were frequent throughout the thirties.

16. The Curtis man was a sales representative of the Curtis Publishing Company, the Philadelphia-based publishing giant. The firm's leading publications in the 1930s included **The Country Gentleman, The Ladies' Home Journal, The Saturday Evening Post,** and **Jack and Jill.**

17. In addition to the emergency light hog and pregnant sow buying program, the AAA also began an emergency corn-loan program in the fall of 1933. This was prompted in part by a decline in corn prices and the realization that the corn-hog program checks would not reach farmers soon enough to tide them over the winter of 1933-1934. Farmers were allowed to borrow 45c per bushel on corn that was sealed in cribs. The loans could be redeemed by sale of the corn at the farmer's discretion. Benedict, **Farm Policies,** p. 308.

18. A tax on the first step of processing financed benefit payments under the initial AAA programs.

19. Earl May—a Shenandoah, Iowa, seed entrepreneur—owned and operated radio station KMA. The station was popular with farmers because of its extensive coverage of agriculture.

20. The highly complicated procedures involved in drawing up thousands of individual corn-hog contracts were dealt with through temporary local committees. County agents, part of the county extension service, assisted these committees. The agents had been trained in December 1933 after the AAA contract forms and administrative rulings were published. Local farmers made up these temporary county and township level committees; Elmer served on one such committee in Amaqua Township. In January 1934 the county agents held two-day training sessions for the township

committees. Committeemen filled out and signed contracts themselves during the training sessions. By the time the actual sign-up campaign was launched, there were between 125 and 150 trained men in each county able to take the contracts to the farmers, help interpret the documents, and assist in filling out production information which would be the basis of the support payments. Roberts, "Corn-Hog Program in Iowa," pp. 25-30; Joseph S. Davis, On Agricultural Policy, 1926-1938 (Palo Alto: Stanford Univ. Food Res. Inst., 1939), pp. 247-71; Des Moines Register, Jan. 10, 21, 1934.

21. Under the corn-hog program farmers had to substantiate with "supporting evidence" (documentation) their past production and disposal of hogs. This was time consuming and complicated but provided the basis for calculating payments.

22. Rather than create an entirely new bureaucracy, federal administrators used the Farm Bureau Federation as an administrative arm of the AAA. The reliance on the Farm Bureau did nothing to make the AAA more popular with the already dissident Holiday Movement and Farmers Union. For the role of the Bureau, see Christiana McFadyen Campbell, The Farm Bureau and the New Deal: A Study in the Making of National Farm Policy (Urbana: Univ. Ill. Press, 1962), pp. 66-67.

23. The final figures in Iowa were 175,765 contracts signed and an estimated $73 million paid in benefits. Roberts, "Corn-Hog Program in Iowa," p. 40.

24. To ensure local control of the program, county control associations were formed to exercise authority over allotments and all matters pertaining to the corn-hog program. Farmers who signed contracts automatically became members of the association. Permanent committees were chosen on the township level by secret ballot. The committeemen were paid and held powerful positions since they measured the contracted corn acres and estimated yields. These estimates served as the basis for payments and so were vital to farmers in the program. Ibid., pp. 42-46.

25. Governor Clyde Herring called a special session of the Iowa legislature which met from November 1933 to March 1934. The most significant, and controversial, legislation produced by the extra session was the "Three-Point Tax Replacement Law." The new statute included an income tax, a corporation tax, and a sales tax. See Leland L. Sage, A History of Iowa (Ames: Iowa State Univ. Press, 1974), pp. 301-2.

26. One of FDR's earliest relief efforts was to reorganize farm credit institutions. In March 1933 he abolished by executive order the old Farm Board and set up the Farm Credit Administration, which coordinated what had once been a piecemeal system. The immediate result was to ease the pressure on farm mortgages by freeing credit for farmers. The Federal Land Bank was also organized under the Farm Credit Administration. The Farm Credit Act of 1933 carried the consolidation further and set up a system of land banks, credit banks (for production credit), and banks that would loan funds to cooperatives. Twelve production credit corporations were established, with a nationwide system of local production credit associations. One of the production credit associations could have supplied production funds to Elmer, but there would have been considerable delay.

It is interesting to note that the signing of a corn-hog contract appears to have destroyed a farmer's local credit. See Benedict, Farm Policies, pp. 280-83; First Annual Report of the Farm Credit Association (Washington: USGPO, 1934); and especially Earl L. Butz, The Production Credit System for Farmers (Washington: Brookings Inst. 1944).

27. Ledges State Park, located on the Des Moines River, helped to serve the recreational needs of central Iowans. The park was approximately fifteen miles southeast of the Powers farm.

28. Elmer referred to the land taken out of production under the corn-hog program as "contracted acres."

29. Delays in the corn-hog program justifiably disillusioned Iowa farmers. Iowans had moved quickly through the sign-up process, leading the country in dealing efficiently with the complicated procedures. Problems soon developed, however. The difficulty was the checking of acreages and hog-sales figures that were to be used as the basis for payments. A complex system of county and state quotas was issued, based on statistical analysis of past production, and corn-hog contracts were then adjusted downward to bring farmers' estimates into harmony with the statistical analysis. For the most part, Iowa's figures were extremely accurate, and only a small amount of downward adjustment was necessary. Yet, the procedure was time consuming, and when coupled with the spring drought it caused farmers who had signed contracts to chafe. Further delays occurred during the summer. Many Iowa contracts, adjusted and reviewed, were in Washington, D.C., ready for acceptance and the issuance of the first checks by July. However, the speed of the Iowa operation and the low level of downward adjustment in base estimates caused a political protest from other states whose work was progressing more slowly and whose estimates were trimmed more radically. The fact that Secretary Wallace and Albert G. Black (head of the AAA Corn-Hog Section) were both Iowans raised suspicions in other states. Wallace was forced to delay the corn-hog checks for the Hawkeye state while an investigation was carried out. Checkers from other states were sent into Iowa to make their own estimates of acreages and hog figures. After this was complete and few discrepancies found, checks were finally issued in the fall of 1934. By this time one of the major problems of the mid-1930s was apparent: government-planned production cutbacks were to coincide with drought. The result was almost total disaster. Government planning and bad weather conspired to drive the food supply from surplus to scarcity. Roberts, "Corn-Hog Program in Iowa," pp. 85-115; Wallaces' Farmer 59(Aug. 18, 1934):472.

30. Rexford Guy Tugwell, a former economics professor at Columbia University, was Assistant Secretary of Agriculture in 1934. He later became head of the Resettlement Administration. A leading study of this member of FDR's "Brain Trust" is Bernard Sternsher, Rexford Tugwell and the New Deal (New Brunswick: Rutgers Univ. Press, 1964).

31. Elmer used two types of horse-drawn plows, a multibottom rig pulled by several teams, and a sulky plow, a one-bottom affair drawn by a single team. DLP Interview.

32. The Democrats continued their successes in the off-year election of 1934. In

Iowa, Democratic Governor Clyde Herring defeated GOP opponent Dan Turner by 74,287 votes, an increase of almost 20,000 over his margin of victory in 1932. The Democrats also retained control of all the statehouse offices. Voters across the state sent six Democrats and three Republicans to the New Deal Congress; Elmer's representative, Republican Fred C. Gilchrist, survived. In Boone County, Republicans held a majority for governor and most of the state offices, but in Elmer's own township the Democrats retained their plurality. Herring defeated Turner in Amaqua Township, 71 to 54, and with the exception of Gilchrist, all the Democratic candidates won in Amaqua. **Iowa Official Register, 1935-36,** pp. 290-91; **Official Congressional Directory, 1935,** pp. 33-35.

33. Al is Al Boles, a passing hired man.

CHAPTER 3: OWNER-OPERATOR: 1935

1. Not only did cattle prices rise markedly during 1934 but Iowa hog prices rose from $3.06 to $4.15 per hundredweight, and corn increased from 43.9¢ to 85.3¢ per bushel during this twelve-month period. See Lauren K. Soth, **Agricultural Economic Basebook of Iowa** (Ames: Iowa State Coll., 1936), pp. 16-20.

2. Anticipating ownership of his farm, Elmer applied to the Federal Land Bank (part of the Farm Credit Administration) for a mortgage-relief loan. The purpose of the Land Bank was to provide easy, long-term credit for farm owners so they could retain ownership during the Depression. All of the land left to his sons by Samuel Powers was encumbered with debt. In the case of Elmer's farm an insurance company held the mortgage.

3. Despite doubts Elmer did sign a new corn-hog contract for 1935.

4. John Thompson, associate editor of **Wallaces' Farmer** since 1932, edited **The Iowa Homestead** from 1918 to 1929. The two Iowa farm periodicals combined in October 1929 and Thompson temporarily retired from agricultural journalism.

5. Huey Long, "the Kingfish," was a flamboyant, perhaps demagogic, politician from Louisiana. Originally a supporter of the New Deal and FDR, he broke with the administration and launched a campaign for a more radical solution to the Depression. He carried his "Share the Wealth" message to Iowa in April 1935. See the **Des Moines Register,** Apr. 28, 29, 1935.

6. The only occasion when Elmer had a good word for the federal public works relief program came when he benefited by free labor for a new outhouse. In truth, by 1935 the FERA conducted projects far removed from the "leaf-raking" or busy work that characterized the earlier efforts of the CWA. When the FERA work program ended in late summer 1935, it had completed nearly 240,000 projects and had given employment to a monthly average of two million Americans. See **The Emergency Work Relief Program of the FERA** (Washington: FERA, 1935).

7. The U.S. Supreme Court in a unanimous decision on May 27, 1935, struck down the National Industrial Recovery Act and its administrative agency, the NRA (symbolized by the "Blue Eagle").

8. Elmer had run afoul of the Iowa law requiring commercial carriers to buy special permits. He had purchased only the usual license plates for his Ford truck.
9. On October 3, 1935 the Italian Army invaded Ethiopia.
10. See note 27, Chapter 1.
11. In preparing to borrow on corn and oats sealed in cribs, Elmer became involved with yet another of the New Deal's "alphabet agencies," the Commodity Credit Corporation (CCC). Stimulated by requests from Corn Belt governors, the Roosevelt administration devised a program to allow farmers to store surplus commodities until they could be profitably sold or until needed on the market. Created in October 1933, the CCC was a cooperative venture of the AAA, the Farm Credit Administration, and the Reconstruction Finance Corporation. Farmers who had signed AAA reduction contracts were eligible for 4 percent loans on stored and sealed commodities. The farmer could pay off the loan at any time and repossess the commodities, or he could turn them over to the CCC at the loan rate. The most radical departure from the past was the loaning of funds on sealed grain in cribs. Initially, corn was allowed a loan of 45 cents per bushel, and oats 12 cents. See Benedict, **Farm Policies, pp. 332-33.**
12. Elmer's local Debt Advisory Committee was part of a voluntary state and county debt adjustment committee system which had been created during the agricultural crisis of 1933. At first, the Farm Credit Administration supervised the debt adjustment committees; however, the Resettlement Administration took over the program in September 1935. See **First Annual Report: Resettlement Administration** (Washington: USGPO, 1936).
13. Perry, located sixteen miles south of the Powers farm, served as an important trading center for portions of Boone, Dallas, and Greene counties.

CHAPTER 4: DRIFTS, DEBTS, AND DROUGHT: 1936

1. On January 6, the Supreme Court in a split decision ruled portions of the AAA unconstitutional. The Court thereby knocked out a second major New Deal program in the space of two weeks. Although the Department of Agriculture was allowed to continue surplus removal, commodity loans, and marketing agreements, it could no longer enter into contracts with producers nor could it continue to levy the processing tax that had paid for reduction programs. New measures to circumvent the Court's decision and to continue direct aid to farmers were soon devised, however. Charles S. Collier, "Judicial Bootstraps and the General Welfare Clause: The AAA Opinion," **George Washington Law Rev.** 4(Jan. 1936):211-42; Benedict, **Farm Policies,** p. 348.
2. Despite its defects and local farmers' complaints, the AAA had been relatively effective in its main tasks. Farmers had received cash to see them through the emergency, and production had been reduced. The drought of 1934 had done as much as the government to cut back production, but whatever the cause, prices were much higher in 1936 than in 1931-

32. In fact, nationwide prices were at 90 percent of the parity years of 1909-14.

3. This was U.S. Highway 30, better known as the "Lincoln Highway," located one mile south of the Powers farm.

4. The previous day marked the "remilitarization" of the Rhineland by Hitler's army.

5. On February 29 Congress, pushed by Secretary Wallace, passed the Soil Conservation and Domestic Allotment Act, a substitute for the AAA. Under the guise of soil conservation, the new act's main objective was raising the level of per capita farm income. The measure offered farmers cash bounties for planting soil-enriching grasses and legumes instead of soil-depleting commercial crops. Farmers were to be paid from congressional appropriations, not processing taxes. The effort was also aimed at a longer term solution to price imbalance through conservation, a lesson learned from the drought years. Benedict, **Farm Policies,** pp. 349-52; Theodore Salutos, "New Deal Agricultural Policy: An Evaluation," **J. Am. Hist.** 61(Sept. 1974):397; Schapsmeier and Schapsmeier, **Henry A. Wallace,** pp. 216-18.

6. **Capper's Weekly,** a rural news and opinion magazine, was published by the well-known Kansan, Arthur Capper of Topeka, one-time U.S. senator from that state. See Homer E. Socolofsky, **Arthur Capper** (Lawrence: Univ. Kans. Press, 1962).

7. The third party was the Union Party, which in Iowa was a mostly agrarian-based political movement that sought a national inflationary fiscal policy on the part of the federal government, easier mortgage refinancing, and guaranteed income for farmers equal to their production costs. The party's candidate for president was William Lemke of North Dakota. He was supported by Father Charles Coughlin and Dr. Francis Townsend, both champions of radical solutions to the Depression and bitter foes of the New Deal. See David H. Bennett, **Demagogues in the Depression: American Radicals and the Union Party, 1932-1936** (New Brunswick: Rutgers Univ. Press, 1969); Edward C. Blackorby, **Prairie Rebel: The Public Life of William Lemke** (Lincoln: Univ. Nebr. Press, 1963); and Donald R. McCoy, **Angry Voices: Left of Center Politics in the New Deal Era** (Lawrence: Univ. Kans. Press, 1958).

8. The Chicago & North Western Railroad operated a large roundhouse and repair facility in Boone. The town, too, was headquarters for the railroad's Iowa Division.

9. The Quaker Oats Company began its ownership of elevators in Iowa in 1912. By 1933 the firm owned nearly sixty elevators in the central states and a score in western Canada. The company's largest processing plant in the 1930s was located in Cedar Rapids, Iowa. See Harrison John Thornton, **The History of the Quaker Oats Company** (Chicago: Univ. Chicago Press, 1933), pp. 230-52.

10. Governor Alfred M. Landon of Kansas was Roosevelt's Republican opponent for the presidency in 1936.

11. The city visitors were relatives of both Minnie and Elmer. They worked for

the Maytag Company in Newton, Iowa, and for the John Deere tractor plant in Waterloo, Iowa. **DLP Interview.**

12. President Franklin Roosevelt, following a tour of the "Dust Bowl," met with midwestern governors and his November opponent, Landon, in the office of Iowa Governor Clyde Herring to discuss drought problems. **New York Times,** Sept. 3, 4, 1936; Michael W. Schuyler, Agricultural Relief Activities of the Federal Government in the Middle West, 1933-1936, unpubl. doctoral diss., Univ. Kans., 1969, pp. 310-12.

13. Movie bank nights, usually held on Saturdays, consisted of drawings for cash prizes to be awarded to members of the audience. The habit of frequent movie-going, which had developed during the late 1920s, disappeared during the Depression; theater managers had to work hard to attract patrons and therefore turned to gimmickry. See Forbes Parkhill, "Bank Night Tonight," **The Saturday Evening Post** 210(Dec. 4, 1937):20-21, 82.

14. Roosevelt and the Democrats again swept Iowa in 1936, but with a reduced margin. Across the state FDR polled 621,756 votes to 487,977 for Landon. Lemke, the Union Party candidate, totaled 29,687. FDR's victory was by 133,770 votes, down from the 183,586 votes which he held over Hoover in 1932. In Amaqua Township, Roosevelt won 129 to 38. **Iowa Official Register, 1937-38,** pp. 288, 313.

15. The "poor farm" was a commonly used name for a county welfare home, which was situated on a farm in most Iowa counties. Destitute county residents—usually elderly or mentally retarded—lived on the farms and raised crops and livestock under the direction of a resident supervisor.

16. Will Rogers, a movie idol to many Americans, starred in several films that pointed out the virtues of small-town farm life. However, his most famous farmer role came in the 1933 film, **State Fair,** based on Phil Stong's 1932 novel of the same name about the Iowa fair. See Andrew Bergman, **We're in the Money: Depression America and Its Films** (New York: Harper & Row, 1972), pp. 71-72.

EPILOGUE

1. **EGP Diary,** Mar. 13, 1941.
2. **EGP Diary,** May 31, 1939.
3. **EGP Diary,** Mar. 24, 1937.
4. **EGP Diary,** Nov. 25, 1942.
5. Elmer's obituary is found in the **Ogden Reporter,** Dec. 31, 1942. The paper duly observed, "Mr. Powers took an active interest in the affairs of the community and he will be greatly missed by his loved ones."

INDEX